Temperature Sensitivity in Insects and Application in Integrated Pest Management

Temperature Sensitivity in Insects and Application in Integrated Pest Management

EDITED BY

Guy J. Hallman
and David L. Denlinger

Routledge
Taylor & Francis Group

LONDON AND NEW YORK

First published 1998 by Westview Press, Inc.

Published 2019 by Routledge
52 Vanderbilt Avenue, New York, NY 10017
2 Park Square, Milton Park, Abingdon, Oxon OX14 4RN

Routledge is an imprint of the Taylor & Francis Group, an informa business

A CIP catalog record for this book is available from the Library of Congress.

ISBN 13: 978-0-367-28983-6 (hbk)
ISBN 13: 978-0-367-30529-1 (pbk)

Contents

Preface

In response to the spiraling economic and environmental costs resulting from an inordinate reliance on chemical pesticides in recent history, integrated pest management (IPM) was developed to provide a sustainable pest control strategy. The goal of IPM is to combine all practical pest mitigation techniques in a harmonious manner that strives to optimize economic costs and minimize environmental degradation. Major effort worldwide is devoted to developing reliable pest management systems, and numerous volumes have been dedicated to the subject.

Insects vary much more in response to temperature than their poikilothermic nature would initially indicate. In terms of behavior, insects orient themselves as needed to benefit from heating and cooling opportunities. Physiologically, insects can withstand freezing temperatures for prolonged periods, and some brave heat that drives other forms of animal life to cover.

Temperature Sensitivity in Insects and Application in Integrated Pest Management concentrates on an array of IPM tools that exploit extreme temperatures, a feature that has received little notice in most other IPM tomes. The biological basis for using temperature extremes in controlling insects is analyzed, and practical IPM techniques that rely on temperature are presented.

Two people who helped in the preparation of this volume deserve special mention: Annette Manzanares typed much of the book, and Aleena M. Tarshis Moreno prepared many figures.

Guy J. Hallman
David L. Denlinger

1

Introduction: Temperature Sensitivity and Integrated Pest Management

Guy J. Hallman and David L. Denlinger

Temperature is one of the principal factors delimitating survival and reproduction of insects and mites. Temperature extremes are a cause of significant natural mortality in populations and offer a rich potential that can be exploited for the development of environmentally safe pest management strategies.

The omnipresence of temperature stress has resulted in a wealth of physiological and behavioral adaptations that have evolved to ameliorate or avoid the full brunt of high or low environmental temperatures. These range from behaviors as simple as moving in or out of sunlight to increase or decrease body temperature to the more complex social behaviors of honey bees, *Apis mellifera*, which cluster to preserve warmth during severe cold and use evaporative cooling aided by wing movement to cool the hive during hot weather. Physiologically insects prepare for cold weather in temperate climates by such means as increasing concentrations of cryoprotectants in the hemolymph and arresting development at a certain cold tolerant stage. A series of proteins produced in response to extreme temperatures and other stresses increases tolerance of the organism to further stress. Some ants are able to initiate heat shock protein synthesis in the absence of thermal stress and use that ability to prepare for brief forays into the desert during the hottest part of the day to scavenge for organisms that have succumbed to the heat (Gehring & Wehner 1995). Coleman et al. (1995) discuss the need for integrating molecular function, metabolic cost, and ecological manifestation and variation to better understand the evolutionary significance of heat shock proteins and their use in managing problems of thermal stress.

Both heat and cold have been used to suppress pests since the beginnings of insect control and have been classified under the broad category of physical controls which Metcalf et al. (1962) defined as methods which employ abiological properties of the environment to the detriment of pests. Prior to the widespread use of fumigants, cold and heat were used to a considerable extent to control pests of a wide array or stored products including grain, bulbs, logs, fruit, and cloth. One of the reasons for burning crop residue was to destroy pests, and fall plowing was often effective in exposing quiescent stages of pests protected in the soil to lethal winter temperatures.

Pest management through temperature manipulation is receiving renewed interest as a non-chemical method which poses no residue problem. Heat and cold received two-thirds of the United States Department of Agriculture, Agricultural Research Service effort to develop quarantine treatments in 1992 (Fig. 1.1). Other countries show similar statistics. For example, all of the effort to develop quarantine treatments for fresh commodities by the Queensland Department of Primary Industries during 1994-1996 concerned heat and cold (Anonymous 1996). Use of methyl bromide, a widely used fumigant for killing insects and plant pathogens in soils, is scheduled to cease within several years as it is considered a significant stratospheric ozone depleter. Its impending loss has spurred research into alternatives, including temperature treatments for soil, fresh commodities, stored products, and structures. Heat, including solarization, electronic heating, and steam, is being studied as a replacement for methyl bromide fumigation of planting beds and containers (Anonymous 1995). Because of consumer concerns with not only methyl bromide but other fumigants as well, heat and cold treatments for wood-boring insects have been developed and are currently being used to treat structures (Chapter 7).

Temperature manipulation of insects is finding its way into pest management programs. Cold storage is used to increase the shelf life of biological control agents and the hosts on which they are reared (Chapter 9). A female temperature-sensitive strain (*tsl*) of Mediterranean fruit fly, *Ceratitis capitata*, is presently being studied in large-scale field trials for suppression of this pest by means of release of sterile males (Economopoulos 1996). Female Mediterranean fruit fly eggs of this strain are more susceptible to heat than males and are selectively killed when both are immersed in heated water. This procedure reduces rearing expenses by almost half because females do not develop beyond the egg stage. In addition, elimination of females prevents fruit perforation damage caused by field releases of large numbers of sterile females.

Heat and cold

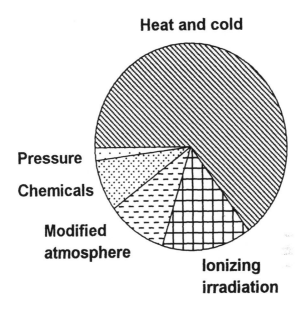

Pressure

Chemicals

Modified atmosphere

Ionizing irradiation

FIGURE 1.1 Proportion of the total research effort in the U.S. Department of Agriculture, Agric. Research Service dedicated to each of five quarantine treatment categories in 1992 (Anonymous 1992).

Today, the successful application of temperature to integrated pest management is enhanced by an increased understanding of the physiology of insect sensitivity to temperature extremes, and especially the effect of variables such as nutrition, humidity, and host, in modifying the reaction. Bale (1996) proposed five ecophysiological strategies of cold-hardiness: freeze tolerance, freeze avoidance, chill tolerance, chill susceptibility, and opportunistic survival, with the goal of integrating a quantitative component of cold hardiness into the overall equation. Being able to more precisely classify cold-hardiness mechanisms and estimate numbers of survivors enables more directed and effective approaches in using cold to manage insect populations. Contrary to a generally accepted hypothesis, contact with the host plant, in this case winter barley, was beneficial, not detrimental, to survival of bird cherry-oat aphid, *Rhopalosiphum padi*, exposed to freezing temperatures (Butts et al. 1997). The relationships between cold hardiness, diapause, and tolerance to desiccation are intriguing, and mechanisms that evolved in response to one form of stress may now offer protection against other stresses (Block 1996, Pullin 1996).

4

Temperature can exert subtle influences on plants that, in turn, influence the success of insect herbivores. Stamp & Yang (1996) discussed interactions between temperature and three tomato allelochemicals on the relative growth rate of three lepidopteran herbivores and discussed the implications for insect outbreaks and pest management. Under certain circumstances the presence of allelochemicals enhanced insect growth instead of retarding it.

The natural world is full of intriguing examples of temperature exploitation. Heat is sometimes used by insects to kill other insects. When attacked by the Japanese giant hornet, *Vespa mandarinia japonica*, hundreds of Japanese honey bees, *Apis cercana japonica*, engulf the hornet in a ball which quickly reaches 47°C, a temperature lethal to the hornet but not the bees (Ono et al. 1995). Bumblebees use the low temperatures prevailing outside the colony to rid themselves of parasites (Müller & Schmid-Hempel 1993).

Finally, a more ominous reason for devoting increased attention to thermal sensitivity is the possible global warming trend which, if it is not simply a transient blip in the global temperature record, may be impossible to avoid, and, thus, will have to be included in future pest management models. Global warming raises the prospect of both increasing and decreasing current ranges of certain pest species. Assessing the evolutionary potential of insects to respond to temperature change will thus be critical for predicting the impact of such global shifts.

References

Anonymous. 1992. *Quarantine Workshop for Horticultural Commodities: Final Report.* U. S. Department of Agr., Agric. Research Service. 93 pp.

Anonymous. 1995. *Status of Methyl Bromide Alternatives Research Activities.* Crop Protection Coalition, Fresno, California. 78 pp.

Anonymous. 1996. *Horticulture Postharvest Group Biennial Review 1996.* The State of Queensland, Department of Primary Industries, Brisbane. 42 pp.

Bale, J. S. 1996. Insect cold hardiness: a matter of life and death. European J. Entomol. 93: 369-382.

Block, W. 1996. Cold or drought-the lesser of two evils for terrestrial arthropods? European J. Entomol. 93: 325-339.

Butts, R. A., G. G. Howling, W. Bone, J. S. Bale & R. Harrington. 1997. Contact with the host plant enhances aphid survival at low temperatures. Ecological Entomol. 22: 26-31.

Coleman, J. S., S. A. Heckathorn & R. L. Hallberg. 1995. Heat-shock proteins and thermotolerance: linking molecular and ecological perspectives. Trends in Ecol. and Evol. 10: 305-306.

Economopoulos, A. P. 1996. Quality control and SIT field testing with genetic sexing Mediterranean fruit fly males. *In* McPheron, B. A & G. J. Steck, eds. *Fruit Fly Pests: A World Assessment of Their Biology and Management*, pp 385-389. St. Lucie Press, Delray Beach, Florida.

Gehring, W. J. & R. Wehner. 1995. Heat shock protein synthesis and thermotolerance in *Cataglyphis*, an ant from the Sahara desert. Proc. Nat. Acad. Sci. USA 92: 2994-2998.

Metcalf, C. L., W. P. Flint & R. L. Metcalf. 1962. *Destructive and Useful Insects: Their Habits and Control.* McGraw-Hill, New York.

Müller, C. B. & P. Schmid-Hempel. 1993. Exploitation of cold temperature as defence against parasitoids in bumblebees. Nature 363: 65-67.

Ono, M., T. Igarashi, E. Ono & M. Sasaki. 1995. Unusual thermal defence by a honeybee against mass attack by hornets. Nature 377: 334-336.

Pullin, A. S. 1996. Physiological relationships between insect diapause and cold tolerance: coevolution or coincidence? European J. Entomol. 93: 121-129.

Stamp, N. E. & Y. Yang. 1996. Response of insect herbivores to multiple allelochemicals under different thermal regimes. Ecology 77: 1088-1102.

2

Physiology of Heat Sensitivity

David L. Denlinger and George D. Yocum

Insects are enormously vulnerable to high temperature injury. Heat from the sun or artificial heat can quickly elevate body temperature in these small-bodied poikilotherms to lethal levels. Add to this the challenge of maintaining water balance at high temperature and the problem becomes even more formidable. Exposure to temperatures that are too high can be a daily threat to survival. But, insects have access to an array of behavioral and physiological responses that can be elicited to circumvent or minimize potential injury. As a fly darts from a sunny spot to the shade or a caterpillar crawls from the upper to the lower surface of a leaf body temperature can change many degrees in just a few minutes. Such behaviors represent a first line of defense against high temperature injury. Brief forays into high temperature zones are readily tolerated, as long as the insect has the option of retreating frequently to a more moderate environment to prevent overheating. As a second line of defense, insects have a fascinating suite of physiological and biochemical adaptations that help to prevent injury caused by thermal stress. And, even though they are poikilotherms, insects are not completely at the mercy of their thermal environment. Basking or active exploits such as shivering can be used to boost body temperature, while some insects "sweat" and use evaporative cooling to lower their body temperature. At the cellular level, high temperature survival is enhanced by the synthesis of stress proteins and other key metabolites.

While the deleterious effects of high temperature are most obvious, insects may, in certain cases, exploit high temperatures for their own benefit. Insects infected with viruses, bacteria, protozoans, fungi, and parasitoids frequently seek high temperatures to rid themselves of infection (Heinrich 1993). The infection apparently elicits, among other responses, a behavioral response directing the insect to bask. Elevating body temperature to 35-40°C can help ward off lethal infections. High temperatures can also be exploited to escape predation. The silver ant, *Cataglyphis bombycina,* an inhabitant of the Saharan desert, comes out to forage only for a brief interval at midday when temperatures on the desert floor are at their hottest, over 60°C (Wehner et al. 1992). At these extreme temperatures potential lizard predators are forced to remain inactive in their underground burrows. Diurnal patterns of other day-active insects are also possibly dictated by thermal constraints of vertebrate predators.

In this chapter we explore the nature of heat injury and discuss the adaptations that allow insects to tolerate high temperature. The literature of high temperature effects on insects is voluminous, and we offer a mere sampling of the information that is available. Temperature effects are among the most widely studied features of insect development, and for a good overview of the older literature we recommend consulting a comprehensive review by Uvarov (1931). More recent coverage of selected aspects of high temperature insect biology includes reviews of mechanisms of thermo-regulation (Heinrich 1993), evaporative cooling (Prange 1996), biochemical adaptations to temperature extremes (Hochachka & Somero 1984), microclimate considerations (Willmer 1982a), evolution of thermal tolerance (Huey & Kingsolver 1993), and heat shock proteins (Craig 1985, Lindquist 1986, Lindquist & Craig 1988, Parsell & Lindquist 1993, Craig et al. 1993, Morimoto et al. 1994). Our goal for this chapter is to introduce the features that we feel are most relevant to the potential use of high temperature for strategies of insect control. Tools that might disarm the thermotolerance mechanisms of insects are of special interest for integrated pest management. In practical terms, even a modest reduction in thermotolerance has the potential for making heat treatment economically viable. Not only can treatment costs be reduced by such manipulations, but the risk of injury to the associated commodity can be minimized.

Optimal Temperature

Insects survive, perform, and reproduce across a broad range of temperatures, but they do so with varying levels of success at different temperatures. A thermal performance curve (as shown in Fig. 2.1) can be constructed for nearly any quantitative trait such as egg production, developmental rate, metabolic efficiency, or learning ability. The curve delimits the body temperatures at which certain activity can occur (tolerance zone). The low extreme is the critical thermal minimum, and the upper extreme, the critical thermal maximum. Construction of such a curve will clearly demonstrate that any activity has a temperature at which performance is optimal (optimum body temperature). Characteristically the drop in performance above the optimum body temperature is more precipitous at the high end of the temperature scale than at the low end. When given a choice, insects and other poikilotherms readily select temperature conditions that will maximize their performance (Fraenkel & Gunn 1961). For a single individual the parameters of the thermal performance curve may vary for different activities, and like other characteristics, the shape, height, and limits of the curve are subject to change through natural or artificial selection.

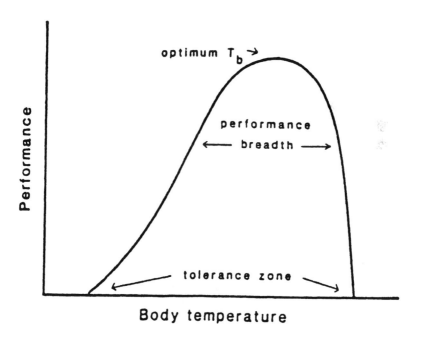

FIGURE 2.1 Hypothetical performance curve for an insect as a function of body temperature (T_b). From Huey & Kingsolver (1989).

Lethal Responses to High Temperature

Brief exposures to very high temperatures can cause immediate death, and it is precisely this type of rapid death that is frequently sought for insect control. Lethality is a function of both temperature and time. The higher the temperature the shorter the exposure needed to kill the insect (Fig. 2.2). The term lethal temperature thus has little meaning unless the time element is also given. At moderately high temperatures survival curves typically have a rather broad shoulder and then drop off rapidly (see curve for 40°C in Fig. 2.2). At higher temperatures (see curve for 45°C in Fig. 2.2), the initial shoulder of the curve is missing and the decline in survivorship is rapid from the onset. A possible explanation for the shoulder is that the insect is capable of surviving a series of non-lethal lesions, but at a certain point, the lesions accumulate to a critical level and cause death. The absence of a shoulder at higher temperatures suggests that lethal lesions develop more quickly under those conditions or that the healing processes that counter the lesions at less severe temperatures are rendered inoperative.

The examples shown in Table 2.1 depict a range of temperatures and exposure times required for lethality. For comparative purposes, such values taken from the literature can be a bit misleading. Not only do investigators use different criteria for lethality, but the statistical treatment of population responses varies. The temperature and time combination needed to kill 50% of a population (LT_{50}) is very different from that needed to kill 99.9968% of a population (probit 9), which in turn differs from the temperature and time required to kill 100% of a population. The only reliable comparative data comes from work using identical experimental and statistical methods.

In addition to the high temperatures that cause rapid death, exposure to a wide range of less severe high temperatures causes thermal wounding that may be manifested at a later stage of development. Such high temperatures are every bit as deadly but the response is not immediately obvious (Fig. 2.3). For example, flesh flies *Sarcophaga crassipalpis* exposed to high temperatures as pupae or pharate adults are killed immediately by a 2 h exposure to 50°C (Denlinger et al. 1991). At a less extreme heat shock, e.g. 45°C for 2 h, the flies survive to complete pharate adult development but then die without escaping from the puparium (Yocum et al. 1994). When heat shocked at 45°C for 80 min, most flies are able to expand their ptilinum and force open the operculum, the cap of the puparium, but they still fail to extricate themselves from the puparium. In such flies the muscular contractions involved in ptilinum expansion are weak, yet the stereotypic behavioral pattern remains

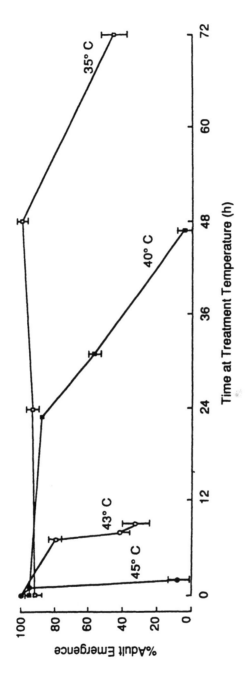

FIGURE 2.2 Survival curves for flesh flies *Sarcophaga crassipalpis* exposed to four different temperatures for various durations. Flies were treated as pharate adults, and survival was based on success of adult emergence. Each point is the mean ±SE of three replicates of 15 flies each. From Yocum & Denlinger (1994).

TABLE 2.1 Examples of temperatures and exposure times required to kill various arthropods

ORDER Family Species	Developmental Stage	Temp. (°C) /Exposure time (min, h=hours)	Definition of mortality	References
ACARI				
Tetranychidae				
Tetranychus urticae	Adult	47/>30	No motion[1]	Cowley et al. (1992)
THYSANOPTERA				
Thripidae				
Heliothrips haemorrhoidalis	Adult	47/10	No motion[1]	Cowley et al. (1992)
HOMOPTERA				
Pseudococcidae				
Pseudococcus longispinus	Adult	47/15	No motion[1]	Cowley et al. (1992)
COLEOPTERA				
Bostrychidae				
Rhyzopertha dominica	Larva	70/5, 80/3	Failure of adult eclosion	Evans (1981)
Curculionidae				
Sitophilus oryzae	Larva	80/3	Failure of adult eclosion	Evans (1981)
	Adult	40/18 h	Not given	Gonen (1977)
S. granarius	Adult	40/24h	Not given	Gonen (1977)

Taxon	Stage	Temperature/exposure	Endpoint	Reference
LEPIDOPTERA				
Pyralidae				
Ephestia elutella	Diapausing larva	40/96h, 43/24h, 45/16h	Failure of adult eclosion	Bell (1983)
Tortricidae				
Epiphyas postvittana	Fifth-instar	47/15	No motion[1]	Cowley et al. (1992)
Cydia pomonella	Embryo	45/40, 46/35, 47/20	Not given	Yokoyama et al. (1991)
	Fifth instar	47/25	Not given	Yokoyama et al. (1991)
DIPTERA				
Tephritidae				
Ceratitis capitata	Embryo	43/155, 45/63	Failure to hatch	Moss & Jang (1991)
	Embryo	46/31, 47/9	Failure to hatch	Jang (1986)
	First instar	45/49, 46/27, 47/19, 48/10	Survive <24h	Jang (1986)
Anastrepha obliqua	Third instar	46.1/76	Failure to pupariate	Sharp & Picho-Martinez (1990)
A. fraterculus	Third instar	46.1/113	Failure to pupariate	Sharp & Picho-Martinez (1990)
	Third instar	46.1/76	Failure to pupariate	Sharp & Picho-Martinez (1990)

(continues)

TABLE 2.1 (continued)

ORDER Family Species	Developmental Stage	Temp. (°C) /Exposure time (min, h=hours)	Definition of mortality	References
A. distincta	Third instar	46.1/66	Failure to pupariate	Sharp & Picho Martinez (1990)
A. ludens	Larva	40.5/960	Failure to pupariate	Darby & Kapp (1933)
	Pupa	40.5/780	Failure of adult eclosion	Darby & Kapp (1933)
	Adult	40.5/180	No motion[1]	Darby & Kapp (1933)
A. suspensa	Embryo	46.4±0.3/62	Failure to pupariate	Sharp (1986)
Bactrocera dorsalis	Embryo	43/213	Failure to hatch	Jang (1986)
		44/49		
		45/12		
		46/9		
	First instar	43/108	Survive <24h	Jang (1986)
		44/76		
		45/52		
		46/25		
		47/16		
	Early third instar	43/88	Failure of adult eclosion	Jang (1991)
		44/56		
		45/50		
		46/39		
		47/22		
		48/19		
	Late third instar	43/85	Failure of adult eclosion	Jang (1991)

B. cucurbitae	Embryo	44/73 44/48 46/35 47/22 48/21	Failure to hatch	Jang (1986)
	First instar	45/42 46/21 47/9 48/5	Survive <24h	Jang (1986)
Sarcophagidae				
Blaesoxipha plinthopyga	Pharate adult	45/180	Failure of adult eclosion	Chen et al. (1990)
Peckia abnormis	Pharate adult	45/120	Failure of adult eclosion	Chen et al. (1990)
Sarcophaga bullata	Pharate adult	45/180	Failure of adult eclosion	Chen et al. (1990)
S. crassipalpis	Third instar	45/30	Failure of adult eclosion	Chen et al. (1990)
	Pupa	45/50	Failure of adult eclosion	Chen et al. (1990)
	Pharate adult	45/120	Failure of adult eclosion	Chen et al. (1990)
	Adult	45/80	No motion[1]	Chen et al. (1990)
Sarcodexia sternodontis	Pharate adult	45/90	Failure of adult eclosion	Chen et al. (1990)

[1]No response to mechanical stimulation.

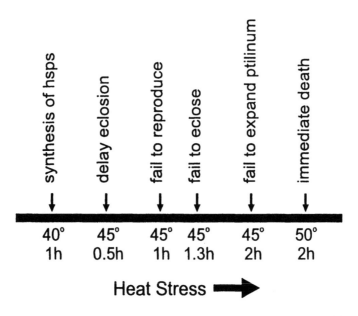

FIGURE 2.3 Symptoms of heat stress generated with different intensities of heat stress in the flesh fly, *Sarcophaga crassipalpis*. Adapted from Joplin & Denlinger (1990), Denlinger et al. (1991), and Yocum et al. (1994).

intact (Fig. 2.4). In this instance the muscular system appears to be more vulnerable to injury than the nervous system generating the response pattern. Flies exposed to 45°C for 60 min emerge successfully from the puparium and the adults survive quite well, but they fail to reproduce. Males are most susceptible to this form of injury. Though they copulate normally, the males fail to inseminate the females (unpublished results). At still less severe conditions (45°C for 30 min) adults are reproductively functional, but they emerge a day late from the puparium and they do so at the wrong time of day! Instead of emerging at dawn, the adults delay eclosion until mid-day. In *S. crassipalpis*, this disturbance of the circadian pattern of adult eclosion is the most subtle overt effect we have noted in response to heat shock, but certainly many biochemical alterations (e.g. shut down of normal protein synthesis and production of heat shock proteins) can be detected with less severe temperature shocks.

We would normally assume that an increase in the stringency of stress conditions would elicit a more robust response. This is usually the case, but not always. A brief exposure to 36°C protects the flesh fly *S. crassipalpis* from subsequent cold shock injury, but temperatures slightly higher or lower

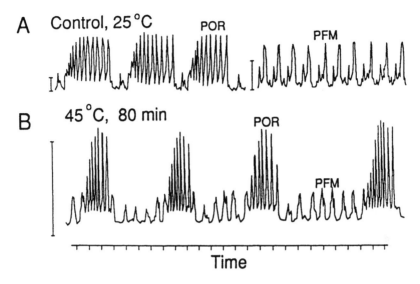

FIGURE 2.4 Representative tensiometric records of ptilinum movements of eclosing adults of the flesh fly *Sarcophaga crassipalpis* that were held at (A) 25°C or (B) were heat shocked at 45°C for 80 min as pharate adults. The time scale indicates 10 s intervals; vertical bars indicate an 0.1 mm displacement of the tensiometric sensor. POR, Program for obstacle removal, a sterotypic behavior program used for removal of the cap of the puparium and for the removal of obstacles; PFM, Program for forward movement, a stereotypic behavior program used to move forward when unobstructed by obstacles. Note that the pattern remains intact in heat-shocked flies, but the intensity of the muscle contraction is greatly diminished. From Yocum et al. (1994).

do not offer this sort of protection (Chen et al. 1987). And, even conditions causing death sometimes defy expectation. While males of *D. melanogaster* are killed by a 4.25-4.5 h exposure to 37.5°C, the same duration at 38°C is reported not to be lethal (Milkman 1962).

Developmental Defects Caused by High Temperature

Heat shock can also elicit interesting developmental abnormalities. These developmental alterations, known as phenocopies, frequently resemble known mutations. Numerous phenocopies caused by heat shock during embryogenesis or metamorphosis have been reported in *D. melanogaster*.

The bithorax phenocopy, first observed by Hadorn (1955), can be induced by a 4-h heat shock to the embryo. A rich variety of phenocopies having aberrant adult bristle shapes and colors or deformities of wing shape or venation can be induced by brief heat shocks administered to pupae or pharate adults (Milkman 1962, Mitchell & Lipps 1978). Which defect is observed is a function of the age of the fly at the time of exposure. As shown in Fig. 2.5, a heat shock of 40.2°C for 40 min causes quite different phenocopies when administered at different times. The sensitive period for the production of each phenocopy is very brief, probably less than 2 h. The various phenotypes are generated by disruption of a heat-sensitive developmental process that is unique to a particular developmental window. For example, the adult fly's bristles are normally black, but the blond-bristle phenocopy can be produced during a brief developmental interval by a heat-shock induced shut-down of phenol oxidase, the enzyme needed for melanin production (Mitchell 1966).

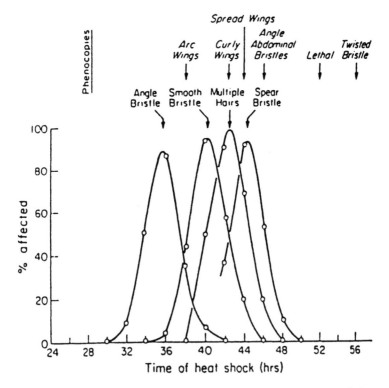

FIGURE 2.5 Frequency of various phenocopies resulting from a 40 min heat shock at 40.2°C in *Drosophila melanogaster* at different times after puparium formation. Data are given for the four characters indicated by curves, and sensitive periods for six additional characters are indicated by arrows. From Mitchell & Lipps (1978).

Several species of *Aedes* mosquitoes can be feminized when reared at high temperatures (Anderson & Horsfall 1963, Horsfall et al. 1964). While male larvae reared at 30°C develop into normal adults, elevating the temperature to 31°C during the last half of the fourth larval instar yields males with female-type antennae, palpi, and oral stylets. Such feminized males do not swarm, copulate nor produce sperm. While a phenocopy defect of this sort obviously renders the mosquito reproductively unfit, many of the phenocopies in *Drosophila* that merely represent alteration of bristle formation or wing venation are perfectly capable of reproducing.

The temperature ranges that elicit phenocopy production are frequently rather narrow, and the transition is abrupt. In *D. melanogaster* the heat shock that causes a phenocopy defect in wing venation is effective only within a narrow temperature window of 39.5 to 41.5°(Milkman 1962). Temperatures below this range do not elicit the effect and higher temperatures usually cause death. The switch to feminization in *Aedes* mosquitoes also occurs very abruptly. At 30°C the palpi of the males are only slightly shorter (feminine), but at 31°C the palps and other structures throughout the body are dramatically feminized (Horsfall et al. 1964).

Variation in Tolerance

There are some remarkable instances of insects tolerating extremely high temperatures. Thermophilic ants that come out to forage only at midday when surface temperatures are above 60°C are well represented in hot, dry deserts around the world (Heinrich 1993). The firebrat thrives in the blistering heat of furnace rooms, and brine fly larvae flourish in hot, saline pools. While the thermophilic species thrive at unusually high temperatures, much more moderate temperatures can prove lethal to black fly larvae or caddisflies that normally inhabit cold mountain streams. Lethal temperature varies both with species and the developmental status of the species, as is evident from the examples shown in Table 4.1.

Selection experiments suggest a genetic basis for heat tolerance, e.g., heat tolerance in *D. melanogaster* can be increased by laboratory selection (Huey et al. 1991, Huey & Kingsolver 1993, Cavicchi et al. 1995, Hoffmann et al. 1997). Yet, wild populations of *D. melanogaster* and *D. simulans* from temperate (38°S) and tropical (17°S) Australia show very little geographic variation in this trait (Hoffmann & Watson 1993), and several species of flesh flies from the tropical lowlands of Panama are even less heat tolerant than closely related species from temperate North America (Chen et al. 1990). The fact that the tropical flies in these studies are not more heat tolerant than the temperate ones is perhaps not surprising because maximum summer temperatures in the temperate zone are not that different from maximum temperatures recorded at the tropical site. Increased heat tolerance is sometimes acquired through selection not directly associated with temperature. A strain of the red flour beetle, *Tribolium castaneum*, selected for resistance to malathion, exhibited increased tolerance to heat stress, but no increase in resistance to stresses caused by cold, desiccation, or starvation (Shukla et al. 1989). Interest in genetic selection of insects for thermal tolerance has recently been renewed as researchers seek model systems to investigate the impact of global warming (Huey & Kingsolver 1993).

A simple point mutation substituting one amino acid for another can have a tremendous impact on the thermal constraints of an organism. For example, one of the acetylcholinesterase mutants in *D. melanogaster*, Ace^{IJ40}, is a conditional mutation that is lethal if they fly is reared at temperatures above $29°C$ (Greenspan et al. 1980). The basis for this heat-sensitive phenotype is the substitution of a single leucine for a proline at one particular site (Mutero et al. 1994). The consequence is a misfolding of the enzyme, a problem that is exacerbated at high temperatures. The wealth of temperature-sensitive, conditional mutations reported in *D. melanogaster* attests to the capacity for a small change to have a major effect on an insect's thermal characteristics.

Within a single species temperature tolerance varies considerably depending on developmental stage. In the flesh fly *S. crassipalpis*, pharate adults were most tolerant of a $45°C$ heat shock, followed by pupae > adults > wandering phase third instar larvae > feeding phase third instar larvae (Chen et al. 1991). Diapause status also plays an important role (Denlinger et al. 1988): the low metabolic activity and cell cycle arrest characteristic of diapause makes most diapausing stages quite tolerant of heat and other forms of environmental stress. Nondiapausing larvae of *Ephestia elutella* exposed to high temperatures develop into sterile adults, but diapausing larvae exposed to the same temperature are not affected (Bell 1983). Even body size may be important. Smaller adults of *Drosophila willistoni* and *D. melanogaster* are killed more quickly by high temperature than larger flies (Levins 1969), possibly due to their greater vulnerability to desiccation. The amount of interexperimental variability commonly observed in stress experiments also suggests that the insect's overall physiological status and state of health contribute significantly to temperature tolerance.

And, temperature tolerance is likely to vary for different tissues within a single organism. Certainly the temperatures experienced by different regions of the body at any one time may be dramatically different. The massive flight muscles in the thorax of adult insects generate huge amounts of heat during flight or shivering, and thus in a flying insect, thorax temperatures may be quite a bit higher than the temperature of the head or abdomen. Flesh flies captured at mid-day in the summer have thoracic temperatures approximately 8°C above ambient, while the temperature in the abdomen is elevated only a couple degrees (Willmer 1982b). Honey bees can remain in flight at air temperatures up to 46°C. At these high temperatures, the temperature of the thorax approximates ambient temperature, but the head is kept cooler by evaporative cooling; fluid regurgitated from the honey crop spreads over the anterior region of the body and cools the head (Heinrich 1980). During pre-flight warm-up thoracic temperature in *Manduca sexta* may jump within a few minutes from 30°C to 45°C, while head temperature increases to 36°C and the abdomen temperature increases to only 32°C (Heinrich & Bartholomew 1971). At an air temperature of 19°C, the thorax of the dragonfly *Anax junius* quickly rises to 32°C during pre-flight warm-up, but the temperature of the abdomen remains below 20°C (Heinrich & Casey 1978).

When exposed to solar radiation, immobilized butterfly wings heat quickly and not uncommonly reach temperatures more than 20°C above ambient (Kingsolver 1985, Heinrich 1993, Schmitz & Wasserthal 1993), while the more massive thorax heats much more slowly and reaches an equilibrium temperature lower than that attained by the wings (Fig. 2.6). Though temperature in the antennae of a papilionid butterfly increases nearly as rapidly as in the wings, the equilibrium temperature attained is much lower, only 4°C above ambient (Fig. 2.6) as opposed to 22°C above ambient for the wings. The butterfly antenna, with its low mass and club-like shape, efficiently disperses radiant heat in all directions and thus minimizes its accumulation. The fact that different tissues are routinely subjected to different levels of heat intensity does not by itself imply differences in tolerance, but we consider such differences to be quite likely. The fact that different tissues within the same organism may synthesize different stress proteins and may have different thresholds for expression (Fittinghoff & Riddiford 1990, Joplin & Denlinger 1990) already suggests fascinating differences in thermotolerance at the tissue level. At the organismal level, the least thermotolerant tissue will obviously be the weak link in survival at high temperatures.

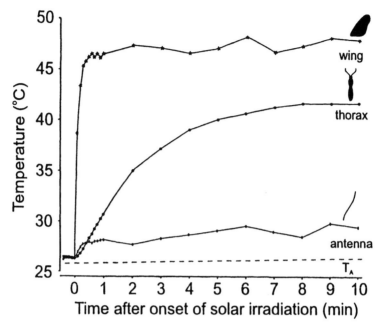

FIGURE 2.6 Warming rates and equilibrium temperatures of wing, thorax and antenna of a papilionid butterfly *Pachliopta aristolochiae* after the onset of solar irradiation (54 Mw/cm²). T_A=Ambient temperature. From Schmitz & Wasserthal (1993).

Causes of Heat Injury

An array of abnormalities is evident at the cellular level in response to heat stress. The microenvironment of the cell (pH and ion concentration), biological molecules (protein, DNA, RNA, lipids, and carbohydrates), and cell structures are all vulnerable to lethal alterations. Additional problems may be inflicted at higher levels of organization, at the level of cell-cell interaction and at the level of the integrative processes that serve to coordinate the higher level physiological processes within the body.

The low molecular weight molecules in the fluid bathing the cells contribute significantly to the cell's microenvironment. These low molecular weight molecules (e.g. various salts, hydrogen ions) influence the charge state of the macromolecular components of the cell (proteins and nucleic acids) and thus can potentially alter the function of the macromolecules and their ability to form cellular structures (Hochachka & Somero 1984). Arthropods maintain a rather constant pH within their normal temperature range, but at high temperatures a drop in pH is common. In the land crab *Stoliczia abbott,* hemolymph pH remains unchanged from 10 to 17°C, but above 17°C, it decreases (Reiber & Birchard 1993). A decrease in hemolymph pH is also observed at high lethal temperatures (above 47°C) in the tenebrionid beetle *Centrioptera muricata* (Ahearn 1970). In the crayfish *Astacus pallipes* high temperature stress causes a drop in hemolymph levels of sodium and an increase in potassium (Bowler 1963). At temperatures above 40°C, hemolymph levels of calcium, potassium, and free amino acids are elevated in the armyworm, *Spodoptera exigua* (Cohen & Patana 1982). In the cerambycid beetle *Morimus funereus* hemolymph concentrations of trehalose drop when winter-collected larvae are exposed to 30°C (Ivanovic 1991).

Heat can alter the intracellular environment as well. Intracellular pH levels drop in mouse mastocytoma P815 cells in response to heating (Yi et al. 1983), and intracellular Ca^{2+} increases in response to heat, apparently in response to heat activation of the inositol 1,4,5-triphosphate Ca^{2+} pathway (Calderwood et al. 1988). Within muscle fibers of the crayfish *Austrotamobius pallipes* thermal stress induces a loss of potassium and an increase in sodium (Gladwell et al. 1975).

Both the quantity and types of macromolecules present in the cell can be altered by heat stress. Perhaps the most conspicuous change is in the pattern of protein synthesis. In response to a sudden increase in temperature the normal pattern of protein synthesis is halted and a new set of proteins, the heat shock proteins, are expressed (see section on mechanisms of thermotolerance). Proteins other than the "classic" heat shock proteins are also expressed during thermal stress. In human HeLa cells fifty new proteins, synthesized from preexisting mRNAs, are expressed when the cells are transferred from 37 to 45°C (Reiter & Penman 1983). Proteins present at the start of thermal stress may undergo conformational changes, some of which are irreversible (Lepock et al. 1987).

Transfer and ribosomal RNAs have complex secondary and tertiary shapes that are essential for function and, like proteins, it is the complexity of shape that renders them susceptible to thermal stress. Loss of conformational integrity can lead to a temporary decrease in function, the dissociation of complexes, or degradation. In the bacterium *Micrococcus cryophilus* heat sensitivity is caused by failure of esterification of certain amino acids with their respective tRNAs (Malcolm 1969). When tRNA from *M. cryophilus* is heated to 30°C conformational changes in the tRNA inhibit the ability of the tRNA to bind amino acids. In Chinese hamster cells, exposure to 43 or 45°C causes a dissociation of the polyribosomes (Arancia et al. 1989). The large 28S ribosomal RNA of animals can be characterized by its response to heat:

protostomian 28C ribosomal RNA breaks into an 18S component whereas the deuterostomian 28C ribosomal RNA degrades with no such component (Ishikawa 1973, 1975).

The ability for DNA to function properly is also affected by heat stress. Heating cells produces lesions in the DNA that become visible as strand breaks only under basic denaturing extraction methods. Two types of these hidden lesions, evident in Chinese hamster cells, can be distinguished based on their rate of induction and repair. One is induced by brief exposure to temperatures in the range of 43-45°C and detected by extraction at pH 13, while the other is produced only after a 30 min exposure to 45°C and extraction at pH 12.2 (Warter & Brizgys 1987). When the cells are returned to 37°C, lesions detected in the extraction at pH 13 are slowly repaired, but the lesions detected in the pH 12.2 extraction remain unchanged. Thermal stress in Chinese hamster cells also induces multipolar mitotic spindles which cause chromosomal misalignment and result in the production of single multinucleated daughter cells that cannot reproduce (Vidair et al. 1993).

Lipids are also highly vulnerable to injury at high temperatures. Lipid peroxidation is evident in guinea pig liver cell membranes, as well as in the microsomes and mitochondria, following heat stress (Ando et al. 1994). In *Schistocerca gregaria* the concentration of cholesterol in flight muscle mitochondria increases when the grasshoppers are placed at 45°C (Downer & Kallapur 1981). Carbohydrates, too, may be affected by high temperature. Glycogen in the fat body is quickly utilized when winter-collected larvae of the beetle *Morimus funereus* are transferred to 30°C, and within 3 days the stores are nearly depleted (Ivanovic 1991). A shift from 30 to 40-45°C in the yeast *Saccharomyces cerevisiae* boosts the trehalose level, a metabolic feature that appears to convey thermotolerance (Attfield 1987, Hottiger et al. 1987).

Alterations in the cell microenvironment and changes in macromolecular conformation and function brought on by high temperature have repercussions for cell structure. Embryos of *D. melanogaster* display a complete collapse of their intermediate filament cytoskeleton following a 30 min heat shock at 37°C (Walter et al. 1990). The intermediate filament proteins, proteins that are maternally derived, aggregate at the nucleus following heat shock rather than forming the cytoskeleton of the cell. Different components of the cytoskeleton vary in their sensitivity to high temperature. In normal rat kidney cells fibers of actin are completely disrupted by exposure to 45°C, whereas the intermediate filaments and the microtubules are less affected (Othsuka et al. 1993). Additional effects of thermal stress on the cytoskeleton are reviewed by Coakley (1987). With the collapse of the cytoskeleton cellular membranes become deformed (Bowler 1987, Lee & Chapman 1987, Laszlo 1992). The cellular membrane and mitochondrial cristae of V79 cells develop pathological lesions in a dose-dependent manner following exposure to 40-45 °C (Arancia et al. 1989).

High temperature exerts a profound effect on the structure and function of macromolecules. The key characteristics of a macromolecule that render it biologically active are its spatial conformation and its ability to change its shape as required to carry out its function. The spatial conformation and flexibility of a macromolecule are determined by several interacting factors: 1) the chemical properties and number of each component part, 2) the number and type of chemical bonds that can be formed intra- and intermolecularly, and 3) the current kinetic energy level of the molecule. Varying any of these factors can result in a change in the macromolecular conformation and therefore in its ability to function. An increase in temperature results in an increase in the kinetic energy of the macromolecule, thereby decreasing the ionic, hydrogen, and van der Waals bonds and increasing hydrophobic interactions of the macromolecule. This, in turn, reduces the ability of the macromolecule to hold its shape or to flex as required to carry out its function (reviews by Alexandrov 1977, Hochacka & Somero 1984, Streffer 1985, Prosser 1986, Jaenicke 1991, Somero 1995). The melting and unfolding of proteins is invoked not only at extremely high temperatures that are ecologically irrelevant, but also at temperatures within physiological ranges (Parsell & Lindquist 1993, Hofman & Somero 1995).

High temperature has the potential to interfere with many constituents of the cell and its environment, and thus many cell processes are potentially vulnerable to injury. But, which of these many dimensions of cell physiology is actually the weak link? At one time or another proteins, nucleic acids, and lipids have all been suggested as the primary site of wounding in thermal death (Roti Roti 1982). Currently, several competing models have been proposed to account for thermal death. A model proposed by Bowler (1987) suggests the plasma membrane as the primary site of thermal wounding. In this model, the plasma membrane is disrupted by thermal stress, and this in turn sets in motion a cascade of events involving the inactivation of membrane proteins, leakage of K^+ out of the cell, and movement of Ca^{2+} and Na^+ into the cell. This is followed by secondary lesions that involve loss of the cell's bioelectrical properties and manifestations of a disturbance in Ca^{2+} balance, leading to inappropriate activation of phospholipases, proteinases, and protein kinases. In turn, these lesions lead to the formation of tertiary lesions involving the breakdown of cell metabolism, loss of homeostasis, and finally death.

The model for thermal death proposed by Roti Roti (1982) also focuses on the plasma membrane but suggests a different scenario. In this case, plasma membrane disruption is followed by protein denaturation, caused either by changes in the cytoplasm composition or by the direct effect of heat. These denatured proteins then adhere to the chromatin and restrict enzymatic access to DNA. The cell eventually dies as a consequence of an increase in DNA damage. Hochachka & Somero (1984) point out that an enzyme's loss of metabolic function due to temperature-induced conformational changes will occur long before the subunits start to dissociate. The loss of an enzyme's catalytic and regulatory functions can occur at a fairly early stage of thermal stress and thus result in the buildup of toxic compounds and in the reduction of critical substrates. Such changes are all reasonable candidates for the cause of thermal death, and it is not at all clear that any single cellular structure or process can be designated as the cause of death. Clearly many cellular processes are vulnerable and high temperature will adversely affect many aspects of the cell or organism's physiology simultaneously. Species differences and differences in developmental stage are also quite likely to influence the site of lethal thermal wounding.

Yet, we must emphasize that the thermal death of a multicellular organism is not usually the consequence of massive cell death *per se*, but is due instead to the loss or disruption of cells in a certain, critical tissue. Though numerous potentially lethal effects can be observed at the cellular level, many of the abnormalities observed in individual cells may, in fact, be moot. For an organism, lethal wounding may be inflicted at a higher level of organization. The more complex the biological system, the more susceptible it is to high temperature stress. Macromolecules are more resistant to thermal stress than cells, cells are more resistant than tissues, and tissues are more resistant than the whole organism (Ushakov 1964, Prosser 1986). This feature explains the prevalence of the "living dead," those organisms that are still alive but will not survive and reproduce due to thermal injury. For example, tissues dissected from heat-killed crayfish *A. pallipes* are still metabolically active as measured by oxygen uptake (Bowler 1963). And, pharate adults of the flesh fly *S. crassipalpis* exposed to 45°C for 2 h remain metabolically active and continue to develop into adults, but they fail to escape from the puparium and then eventually die, presumably when their metabolic reserves have been depleted (Chen et al. 1990).

Whole organisms normally exhibit a distinct sequence of responses to high temperature. As the goldfish *Carassius auratus* is heated it first becomes hyperexcitable, followed by an increase in spontaneous hyperactivity, loss of coordination and equilibrium, and finally coma (Friedlander et al. 1976). Interestingly, this same behavioral sequence is observed if only the cerebellum is heated. In the goldfish, the inhibitory neuronal functions seem to be the most heat sensitive, followed by fine motor control, then course motor control. The neuromuscular system of the crayfish *Procambrus clarkii* also appears to be among the most heat sensitive (White 1983). In this species, the inhibitory neurons are more sensitive than the excitatory neurons, and synaptic action potentials are more sensitive than axonal action potentials. In nerve-muscle preparations from the frog *Rana temporaria* it is the neuromuscular junction that fails first in response to heat stress, followed by the muscle and then the nerve (Grainger 1973). The neuromuscular system is also highly sensitive to heat injury in pharate adults of the flesh fly *S. crassipalpis* (Yocum et al. 1994). In this case, the central patterning of the muscular contractions associated with adult eclosion appears less sensitive than the muscles themselves (Fig. 2.4). The correct pattern is executed by the muscles, but the contractions generated are very weak and insufficient to permit the adult's escape from the puparium. As body temperatures rise in the migratory locust *Locusta migratoria* the inability to generate the correct flight rhythm is caused by failure of the neural patterning within the central nervous system (Robertson et al. 1995).

In *Rhodnius prolixus* and many other insects, heat stress is well known to delay molting or delay the onset of metamorphosis (Mellanby 1954, Wigglesworth 1955). The timing of molting involves a whole cascade of events that starts with the attainment of a certain critical size. Disruption of feeding or the processes of digestion or assimilation could easily prevent the insect from attaining critical size, but if the size criterion is met molting could still be prevented by disrupting a number of potential steps in the hormonal scheme that regulates molting. Failure to molt suggests an absence of ecdysteroids or failure of the brain to release the prothoracicotropic hormone (PTTH) needed to stimulate ecdysteroid synthesis by the prothoracic gland. In a series of experiments with *R. prolixus*, O'Kasha (1968a,b,c) performed neck ligations at different times before and after the critical period for PTTH release and then exposed the ligated nymphs to 36 °C. Only the nymphs neck ligated after the critical period successfully molted. This suggests that the impairment is brain centered rather than a problem associated with the prothoracic gland or target tissues that respond to ecdysteroids. But, very little experimental work has been done in this area, and some of the information is conflicting (Rauschenbach 1991). For example, the fact that an injection of exogenous ecdysteroids will not cause a heat-stressed nymph of *R. prolixus* to molt (Wigglesworth 1955) suggests that the impairment may be at the level of the ecdysteroid response.

High temperatures can also prove lethal to insects by promoting desiccation. For an insect to maintain water balance, water intake must equal the amount of water lost through excretion and transpiration. Cuticular hydrocarbons provide an impressive barrier for water loss, and due primarily to this system of waterproofing, insects characteristically maintain a rather constant volume of water across a wide range of temperatures (Wharton 1985). But, above a certain temperature, the critical transition temperature (CTT), the rate of water loss increases dramatically. For the flesh fly example shown in Fig. 2.7, the CTT occurs at 30°C. CTT values commonly range

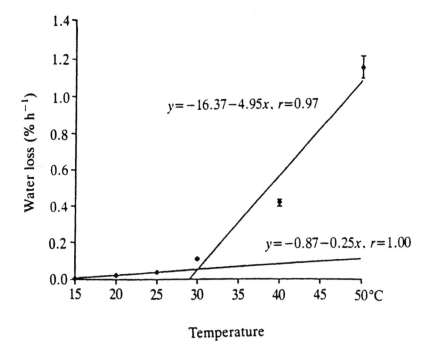

FIGURE 2.7 Rate of water loss in nondiapausing pupae of the flesh fly *Sarcophaga crassipalpis* at different temperatures. The slopes of the lines give the mean activation energy for each temperature range. The intersection of the two lines is the critical transition temperature. Data points represent the mean of 45 pupae; vertical error bars signify 95% confidence limits. From Yoder & Denlinger (1991).

from 30-60°C for different species and developmental stages (Hadley 1994). Although the basis for this dramatic shift in the rate of water loss is still being debated, it is clear that the CTT is dependent on both the quantity and type of hydrocarbons that are present. While the nondiapausing pupae of *S. crassipalpis* shown in Fig. 2.7 have a CTT of 30°C, the CTT for diapausing pupae of the same species is 39°C (Yoder & Denlinger 1991). In this case, the difference between the two is due largely to differences in the quantity of hydrocarbon lining the inner surface of the puparium. The hydrocarbon profiles for the two types of puparia are nearly identical, but the puparium from the diapausing pupa is coated with nearly twice the amount of hydrocarbon found on the puparium of the nondiapausing pupa (Yoder et al. 1992, 1995).

At temperatures above the CTT, the insect will die quickly unless it has access to an abundant supply of drinking water. Water represents 60-70% of body weight for most insects, and many can tolerate a loss of 20-30% of their body water, at least for brief periods (Hadley 1994). The osmotic stress caused by the loss of water and the concurrent increase in solute concentration within the body presumably leads to irreversible cell damage. Direct effects of high temperature are difficult to separate from problems of desiccation unless the insect is maintained at a relative humidity near 100%. Under most field conditions, the effects of these two factors are inexorably linked.

Thermotolerance

Increased thermotolerance can be attained by several routes: genetic adaptation, long-term acclimation, and rapid heat hardening. The potential for genetic adaptation is evident from examples of genetic variation in thermotolerance. When thermotolerance was compared in lines of *D. melanogaster* maintained for 15 years at 18°, 25°, or 28°C, heat shock survival was greatest in the 28°C flies and lowest in the 18°C flies (Cavicchi et al. 1995). The difference in thermotolerance persisted even when the three lines of flies were reared at the same temperature for one generation, thus implying a genetic basis for the differences in thermotolerance. Acclimation capacity can readily be demonstrated by evaluating differences in thermotolerance in individuals exposed for long durations to different temperatures. Numerous experiments indicate greater thermotolerance in insects reared at higher temperatures. For example, when adults of *D. melanogaster* were exposed to 38°C, those reared as larvae at 28°C survived longer than those reared at 25°C, and the 25°C flies, in turn, survived longer than those reared at 18°C (Levins 1969). The third manifestation of increased thermotolerance, rapid heat hardening, can be elicited by a brief exposure to an intermediately high temperature which, in turn, provides protection from injury at a more severe high temperature. It is this response that has been the focus of attention for the huge army of workers who investigate the heat shock response.

Heat shock is the thermal injury caused by a sudden increase in temperature. Heat shock elicits a stereotypic stress response in all sorts of living organisms (Petersen & Mitchell 1985, Parsell & Lindquist 1993, Morimoto et al. 1994). But, this form of injury can be dramatically reduced if the organism is first exposed to an intermediately high temperature. For example, a 2 h exposure of *S. crassipalpis* to 45°C was lethal if the flies were transferred to 45°C directly from 25°C, but if the flies were first exposed to 40°C for 2 h they readily survived a subsequent 2 h exposure to 45° (Fig. 2.8). The actual temperatures and times needed to cause heat shock injury, as well as conditions needed to generate protection, vary from species to species.

In experiments with *S. crassipalpis* reared at 25°C, the optimum temperature condition providing protection against heat shock injury caused by a 2 h exposure to 45°C was obtained by a 2 h exposure to 40°C; temperatures higher or lower than this were less effective (Yocum & Denlinger 1992). Though the protection offered by 40°C developed fully within 2 h, an

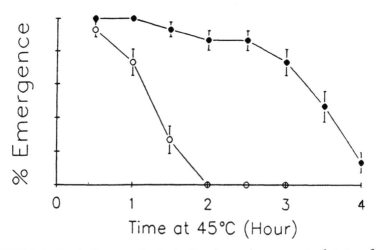

FIGURE 2.8 Survival curves for flesh flies *Sarcophaga crassipalpis* transferred directly from 25°C to 45°C (open circles) or pretreated at 40°C (solid circles). Flies were treated as pharate adults and survival was based on success of adult emergence. Each point is the mean ±SE, n=30-45. From Chen et al. (1990).

increase in thermotolerance was already evident within 30 min. The protection offered by the 2h-40°C pretreatment started immediately after the pretreatment, decreased by 50% within 48 h, and disappeared 72 h later. A 20 min exposure to 40.5°C killed third instar larvae of *D. melanogaster*, but larvae pretreated at temperatures of 31-37°C readily survived 20 min at 40.5°C; temperatures of 33-35°C for 30 min to 1 h offered the greatest protection, while 31 and 37°C were less effective in generating thermotolerance (Mitchell et al. 1979). When adults of *D. melanogaster* were reared at 25°C and then transferred to 29°C for varying periods of time they quickly became tolerant of a 38°C heat shock during the first 12 h at 29°C; up to 9 additional days at 29°C resulted in only a modest additional increase in thermal tolerance (Levins 1969). In the *D. melanogaster* experiments, the thermotolerance generated by a 1-day exposure to 29°C was completely lost 2 days after the flies were returned to 25°C. Consistently, the thermotolerance that protects against heat shock injury is acquired quickly, within minutes, reaches a maximum within a few hours, and then decays rather slowly over several days.

Thermotolerance acquired in response to brief exposure to a moderately high temperature not only protects against death at high temperature but also against phenocopy defects. As discussed previously, the timing of the heat shock is absolutely critical for determining which phenocopy is produced (Fig. 2.5). The hook phenocopy (hook-shaped scutellar bristles), elicited by a 30 min exposure to 41.3°C administered 36 h after pupariation, was prevented when the pupae were first pretreated for 40 min at 35°C (Mitchell et al. 1979). Maximum protection against the phenocopy defect was achieved if the pretreatment was given immediately before the heat shock. A pretreatment 4 h in advance of the critical period for production of the hook phenocopy was not effective in providing protection. A similar protection against the formation of wing vein defects was observed in *D. melanogaster* (Milkman 1962, 1963, Milkman & Hille 1966). Other non-lethal effects of heat shock can also be ameliorated by exposures to moderately high temperatures. The delay in the circadian gate of adult eclosion caused by heat shocking pharate adults of *S. crassipalpis* was avoided by first exposing the pharate adults to a moderately high temperature (Yocum et al. 1994). The acquisition of thermotolerance in *L. migratoria* enabled the locust to continue flying at temperatures 6-7°C above the temperature tolerated by locusts not thermally protected (Robertson et al. 1995).

The obvious advantages of thermotolerance would suggest that, unless there is an associated cost, selection would favor constant activation of the thermotolerance mechanism. But, thermotolerance mechanisms do not function unless a certain threshold has been reached, and they are turned off quickly once the insect returns to a more favorable temperature. This indeed suggests some associated costs. In *D. melanogaster*, the temperature conditions needed to enable adults to survive heat shock conditions exact a toll on fecundity of the females (Krebs & Loeschcke 1994). Other experiments with bacteria suggest that physiological adjustments that increase high temperature survival result in a decrease in competitive ability (Leroi et al. 1994, Hoffmann 1995). High temperatures do elicit a continuum of responses across a wide range (Fig. 2.3), and it is clear that temperatures which generate protection may overlap with temperatures causing deleterious effects. Yet, it is challenging to determine cause and effect, and one cannot assume that a mechanism associated with generation of protection (e.g. synthesis of stress proteins) is necessarily the cause of a reproductive cost. The evidence, at best, remains a correlation.

Several fascinating examples suggest a striking commonality in the responses of organisms to different stressors. This is evidenced by cross protection, i.e., the induction of tolerance to one stressor by exposure to another. A mild heat shock induces tolerance to a normally lethal, low temperature exposure in *S. crassipalpis* (Chen et al. 1987) and *D. melanogaster* (Burton et al. 1988). In the mosquitoes *Anopheles stephensi* and *Aedes aegypti*, a sublethal concentration of the insecticide propoxur induces cross protection against high temperature stress, and conversely high temperature stress protects against injury caused by the insecticide (Patil et al. 1996). Ethanol vapor exposure induces thermotolerance and delays the onset of skin necrosis caused by high temperature exposure in mice (Anderson et al. 1983). Treating the halophilic bacterium *Vibrio parahaemolyticus* at 42°C for 30 min generates not only thermotolerance but also increased tolerance to the heavy metal cadmium and to normally lethal, low osmotic conditions (Koga & Takumi 1995). Pretreating the bacterium *Lactococcus lactis* with low levels of ultraviolet radiation induces protection not only against ultraviolet exposure, but also against ethanol, acid (pH 4.0), hydrogen peroxide and high temperature (Hartke et al. 1995). Such cross protection also can be readily demonstrated in cultured cells: for example, the addition of sodium arsenite or puromycin to cultures of Chinese hamster ovary cells increases thermotolerance (Lee & Dewey 1988). At the organismal level, examples of cross protection are best known from the plant literature (Leshem & Kuiper 1996). Though this type of cross protection has received little attention in insect research, its potential for altering thermotolerance cannot be overlooked.

Mechanisms of Thermotolerance

By far the best known players in thermotolerance are the heat shock proteins. These proteins, which had their scientific founding in F. M. Ritossa's careful documentation of chromosome puffs in salivary glands of D. melanogaster, first attracted attention when they were identified as the protein products resulting from puff activity in heat-shocked salivary glands. This elegant story contributed enormously to our understanding of gene expression. In response to heat stress, the normal pattern of protein synthesis is suppressed, and concurrently several new proteins, the heat shock proteins, are synthesized. These proteins are classified according to their molecular weight, and in D. melanogaster include a high molecular weight protein (82 kDa), members of the 70 kDa family (70 and 68 kDa), and small heat shock proteins with weights of 22, 23,26, and 27 kDa. In S. crassipalpis, the comparable proteins are 92 kDa, members of the 70 kDa family (72 and 65 kDa), and a cluster of small heat shock proteins (23,25 and 30 kDa) (Joplin & Denlinger 1990). The most highly expressed heat shock proteins, members of the heat shock protein 70 (hsp70) family, are highly conserved. The gene for hsp70 is over 50% identical in bacteria and D. melanogaster (Craig 1985). These proteins and their pattern of synthesis in response to heat stress are well documented in all sorts of organisms including bacteria, yeast, plants, insects, fish, and mammals. Humans with a fever synthesize proteins readily recognizable as relatives of the proteins induced by heat stress in E. coli.

But, even within the same organism, slightly different heat shock proteins may be synthesized in different tissues and at different developmental stages. In S. crassipalpis the molecular mass of the major heat shock protein produced in the brain and integument during larval development is a 65 kDa protein (a member of the Drosophila hsp70 family), but at pupariation, synthesis of the 65 kDa protein ceases and thereafter these tissues respond to heat shock by producing a 72 kDa protein (Joplin & Denlinger 1990). In contrast, adult males continue to produce the 65 kDa protein in their terminalia and flight muscles. The embryo appears to be the only stage of fly development that lacks the capacity to synthesize heat shock proteins (Dura 1981). Embryos fail to synthesize heat shock proteins and are also highly vulnerable to thermal injury.

Hsp70 appears to be the most prominent contributor to thermotolerance in insects. This is the protein that responds most dramatically to heat shock, and it can be induced more than 1,000-fold in response to heat shock (Velazquez et al. 1983). The massive boost of synthesis at high temperature suggests a critical role for this protein in survival, yet it is also essential that the heat-inducible hsp70 not be present at normal temperatures. Experiments with cell cultures indicate that the presence of hsp70 at normal temperatures will halt growth and prevent cell division (Feder et al. 1992). In response to heat shock, hsp70 synthesis is rapidly switched on and once the stress has been removed, the protein is quickly eliminated. While the "on" switch is essential for survival at high temperatures, the "off" switch appears just as critical for survival at normal temperatures. Populations of *D. melanogaster* that express elevated levels of hsp 70 in the absence of stress have lower survival at their normal rearing temperature (Krebs & Feder 1997), and overexpression can be detrimental even at high temperatures (Krebs & Feder 1998). Thus, both the timing and level of hsp 70 expression are critical.

Though heat shock was the first stress known to elicit synthesis of these proteins, it is now evident that a number of other forms of stress can prompt synthesis of these same proteins. Arsenite, copper, zinc, and other metals, alcohols, many metabolic poisons (Ashburner & Bonner 1979, Atkinson et al. 1983), accumulation of aberrant proteins (Parsell & Sauer 1989) and even cold shock (Joplin et al. 1990) can elicit the same effect. Thus the term heat shock proteins is a bit of a misnomer, and it is more accurate to refer to these proteins as stress proteins. Even this term is somewhat misleading because many members of these gene families play important roles in cell function at normal temperatures. Yet, the term "heat shock proteins" is so deeply entrenched in the literature and so accurate a descriptor of the response at high temperature that is is likely to persist for years to come.

The actual temperatures that stimulate induction of the heat shock proteins vary in different organisms, but in each case they are induced by temperatures that constitute a stress for that species. Developmental stages that cannot synthesize the heat shock proteins, e.g. early embryonic stages of *Drosophila,* are highly sensitive to heat injury. These features, plus the fact that the proteins are induced so rapidly in response to high temperatures, have for many years provided the basis for the assumption that the heat shock proteins contribute to thermotolerance. More recently, the evidence has moved from correlation to a more firm experimental basis (Parsell & Lindquist 1994). Cultured *D. melanogaster* cells transformed with extra copies of the hsp70 gene acquire thermotolerance more rapidly than normal cells, while cells transformed with hsp70 antisense genes acquire thermotolerance more slowly (Solomon et al. 1991). At the organismal level, recent evidence nicely demonstrates a role for the proteins in thermotolerance. A strain of *D. melanogaster* that carries 12 extra copies of the hsp70 gene acquires thermotolerance more rapidly than the wild type strain (Fig. 2.9).

Not all members of the hsp70 family are heat inducible. Several members of this family of proteins are expressed constituitively, i.e. they are present as normal components of the cell and are not up-regulated in response to heat. In spite of these obvious differences, the heat-inducible hsp70s and the constituitively-expressed hsp70s share a conserved, amino-terminal ATP-binding domain, thus implying a common mechanism for utilizing the energy of ATP.

Heat shock proteins other than those in the hsp70 family have received less scrutiny. The others are usually not as highly expressed as members of the hsp70 family, but they are still a conspicuous element in the high temperature response. Unlike the 1,000-fold increase in expression observed in hsp70, the small hsps in *Drosophila* typically show a more modest 10-fold increase in expression (Arrigo & Landry 1994). And, the temperatures needed for induction are usually lower than those needed for induction of hsp70. The

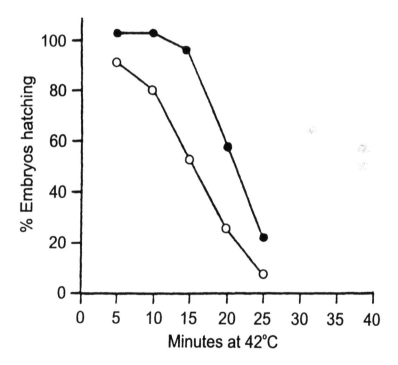

Figure 2.9 Survival curves for wild type embryos of *Drosophila melanogaster* (open circles) exposed to 42°C and embryos that received 12 extra copies of *Hsp70* (solid circles). Embryos were 6 h old at the time of treatment and survival was expressed as the percentage of embryos that hatched. 74-171 eggs were used for each time point. From Welte et al. (1993).

small hsps share a conserved domain, the alpha-crystallin domain, but they are less conserved than the higher molecular weight hsps. All four small heat shock proteins in *Drosophila* respond to high temperature, but they are also expressed at normal temperatures at select times during development. Interestingly, the different proteins have very different expression patterns during development: while hsp27 is strongly expressed during embryogenesis, both hsp27 and hsp23 are expressed highly during pupal and pharate adult development in response to ecdysteroids (Berger & Woodward 1983).

How the heat shock proteins actually protect the organism from injury at high temperature has proven to be an elusive question that has only recently yielded answers. An important clue to the function of hsps in response to heat stress was the observation that many of the diverse stimuli that lead to synthesis of the hsps elicited a common response. The common theme was protein denaturation, and indeed even the injection of denatured proteins induces synthesis of hsps. The initial thought was that hsp70 might somehow be able to recognize denatured intracellular proteins and help restore them to their biologically active shape or identify such proteins for degradation. It is now clear that many functions of members of the hsp70 family are linked to their important role as molecular chaperones (Craig et al. 1993, Parsell & Lindquist 1993, Morimoto et al. 1994). The process of protein folding and assembly is facilitated by molecular chaperones. Molecular chaperones such as hsp70 bind transiently and noncovalently to newly synthesized proteins and to partially unfolded proteins and, by so doing, prevent inappropriate protein-protein interactions and mediate the folding of proteins to their native state. During protein synthesis nascent chains are bound by the molecular chaperones as they are released from the ribosomes. In addition, hsp70s within the mitochondria and endoplasmic reticulum facilitate the translocation of proteins from the cytosol into those organelles by binding with the proteins during early stages of translocation (Ungermann et al. 1994).

Proteins in early stages of folding are particularly vulnerable to high temperature damage, and thus the shut down in normal protein synthesis that is evident at high temperature is a critical aspect of high temperature survival. Properly folded proteins are less susceptible to denaturation and aggregation. But, even mature proteins that are properly folded and localized are vulnerable to extreme temperatures. When present in great abundance, hsp70 and other molecular chaperones can reduce high temperature damage by interacting with the susceptible proteins and thus prevent their interaction with other reactive surfaces. Not all molecular chaperones are heat inducible, but those that are accumulate quickly in response to heat and are available to help maintain the integrity of proteins present in the cell. Some, such as hsp70 and hsp60, appear to be generalists capable of recognizing simple structural motifs shared by many unfolded proteins. Others, such as hsp90, are more specialized in function. Raising the abundance of hsps in response to high temperature should not only help prevent thermal denaturation but also make available a rich supply of chaperones that can be used during the subsequent burst in protein synthesis needed to replace damaged proteins. Since most new protein synthesis is shut down at high temperatures, it is likely that the heat-inducible hsps are functioning mainly to preserve the integrity of proteins that are already present in the cell.

Roles in the synthesis and repair of proteins are not the only function of hsps. Some of the constituitively-expressed stress proteins participate in the regulation of other cellular processes including signal transduction. Hsp90 in mammalian systems is linked to action of the steroid hormones progesterone and glucocorticoids. The stress protein maintains the steroid receptors in an inactive state, and when the hormone is present, it binds to the receptor and releases hsp90. The activated receptor complex is then free to interact with DNA and initiate gene expression (Bresnick et al. 1989). Interestingly, hsp70 also participates in this process (Hutchison et al. 1992) and appears to reassemble hsp90 onto the steroid receptor (Smith & Toft 1993).

The accumulating evidence amply demonstrates the importance of hsps as key elements in the generation of thermotolerance. Yet, it is likely that hsps are only one among many cellular components that contribute to thermotolerance. The fact that hsps do not tell the whole story can be demonstrated in experiments with *S. crassipalpis* (Yocum & Denlinger 1992). The thermotolerance generated in these flies by a 2 h exposure to 40°C decays slowly over a 72 h period. Synthesis of the major hsp in this species, hsp72, stops within the first hour after removal from the high temperature, and the hsp72 is degraded within 24 h. Yet, thermotolerance persists beyond 48 h. This prolonged period of thermotolerance does not appear to be dependent upon the synthesis or persistence of the heat shock proteins. Additional factors are likely to be contributing.

In yeast, heat stress is accompanied by an increase in the disaccharide trehalose (Attfield 1987, Hottiger et al. 1987, 1989), a feature that is thought to promote the reactivation of damaged proteins (Colaco et al. 1992). Most stresses that induce synthesis of the heat shock proteins in yeast also induce an increase in trehalose, with the one exception of canavanine. This amino acid analog induces synthesis of heat shock proteins but has no effect on trehalose. Interestingly, canavanine-treated yeast are not thermotolerant

(Hottiger et al. 1989), thus suggesting the possibility that an elevation in trehalose is essential for the generation of thermotolerance. Henle et al. (1985) tested 16 sugars and sugar analogs for their ability to protect Chinese hamster cells against a normally lethal exposure to 45°C and found that most offered protection. Various small organic and inorganic molecules are likely to contribute to thermotolerance, and indeed many cryoprotectants have the potential to offer protection from heat stress as well as low temperature stress. Classic cryoprotectants such as glycerol and other polyols are reported to protect cultured cells against high temperature stress (Henle 1981, Henle & Warters 1982, Henle et al. 1983, Kim & Lee 1993). While the contributions of such factors to low temperature survival have been examined rather extensively in insects, the possibility that they are also involved in high temperature tolerance has received little attention. The first evidence for polyol involvement, in this case sorbitol, in insect thermotolerance was recently demonstrated in the silverleaf whitefly, *Bemisia argentifolii* (Wolfe et al. 1998).

Heat tolerance is a complex biological response and is not likely to invoke exactly the same responses across the whole range of stress temperatures. This complexity is evident from recent "knock-down" experiments in *D. melanogaster* (Hoffmann et al. 1997). Lines selected for increased knock-down times in response to high temperature did not show lower mortality following heat shock. This suggests that the genes involved in one form of resistance, in this case knock-down resistance, do not necessarily influence other measures of heat resistance. It thus appears that not all dimensions of heat tolerance are linked.

Surprisingly little information is available concerning the role of hormones in eliciting insect stress responses, especially responses to high temperature (Ivanovic & Jankovic-Hladni 1991). The fact that individual cells and cultured tissues can produce heat shock proteins in direct response to high temperature implies that central coordination is not essential for orchestrating that response. Ecdysteroids, however, are well known to stimulate synthesis of small heat shock proteins in *D. melanogaster*, and this event, in turn, correlates with an increase in thermotolerance (Berger & Woodward 1983). Though ecdysteroids are ineffective in promoting thermotolerance in mammalian cells (Chinese hamster ovary cells), the addition of glucocorticoids to the culture medium does increase thermotolerance (Fisher et al. 1986). The possibility for hormonal mediation of other aspects of thermotolerance remains a viable option. There is already good evidence in insects that some biogenic amines such as octopamine (Davenport & Evans 1984) and dopamine (Rauschenbach et al. 1993) function in this manner.

Blockage of Thermotolerance

Stresses that interfere with normal energy metabolism can increase an insect's sensitivity to thermal stress. Inhibitors of oxidative phosphorylation lower thermal tolerance in the Mediterranean fruit fly, *Ceratitis capitata* (Moss & Jang 1991), and hypoxia or anoxia in *C. capitata* (Moss & Jang 1991), the flesh fly *Sarcophaga crassipalpis* (Yocum & Denlinger 1994) the light brown apple moth, *Epiphyas postvittana* (Whiting et al. 1991), and the red flour beetle, *Tribolium castaneum* (Soderstrom et al. 1992) increase sensitivity to high temperatures. The beauty of such treatments is that a much less extreme temperature is lethal. While a 2 h exposure to 40°C would not normally be lethal to *S. crassipalpis*, such an exposure under anoxia proves fatal (Yocum & Denlinger 1994). And, the thermotolerance that is normally generated at 40°C is not acquired under anoxic conditions. Thus, the physiological events needed to attain thermal tolerance require aerobic conditions, and any treatment that denies the organism access to an abundant supply of oxygen should impede the mechanisms of thermotolerance.

Agents capable of reducing thermotolerance have received considerable attention by cancer researchers. By utilizing such drugs to immobilize thermotolerance mechanisms, researchers in this field have been successful in selectively killing cancer cells with high temperature treatments. HeLa S-3 cells cultured under hypoxic conditions with 5-thio-D-glucose lose their ability to become thermotolerant, but 5-thio-D-glucose does not have this effect when the cells are grown in an oxygenated environment (Kim et al. 1978). Pentamidine lowers thermotolerance in glucose-deprived HeLa cells (Kim et al. 1988). Several protein kinase C inhibitors, including tamoxifen and H7, but not HA1004, decrease thermotolerance in a variety of cultured mammalian cells (Mikkelsen et al. 1991). The bioflavinoid quercetin has the peculiar property of being able to block synthesis of the heat shock proteins, and treatment of human colon carcinoma cells with quercetin prevents the acquisition of thermotolerance that is normally acquired by exposure to 42°C (Koishi et al. 1992). Injecting mice with ethanol significantly decreases their ability to tolerate temperature exposures that are normally nonpathological, and it suppresses the thermotolerance normally generated by exposure to moderately high temperatures (Anderson et al. 1983). By analogy, there is likely to be a rich pool of pharmacological agents with similar properties that could be exploited for insect research.

Thermosensitivity

It is evident from our above discussion that an insect's prior thermal history plays a critical role in the development of thermotolerance. While previous exposure to warm temperature is widely appreciated as a stimulant for increasing tolerance to high temperature, it is less well appreciated in the insect literature that, under certain circumstances, exposure to a high temperature can also decrease an insect's ability to survive a future high temperature stress. It is this loss of tolerance that we refer to as thermosensitivity, a term widely used in mammalian cell culture work to refer to this same type of impairment.

Thermosensitivity is exemplified by the response of *S. crassipalpis* to two temporally separated high temperature pulses (Yocum & Denlinger 1993). While pharate adults of this flesh fly survive a 1 h pulse of 45 °C, they die readily if they are subjected to a second pulse 1 day later (Fig. 2.10). Interestingly, even a much more modest second temperature pulse of 35 °C is lethal. Sensitivity to the second thermal challenge slowly decays over a 3 day period. When flies are first made thermotolerant by a 40 °C pretreatment prior to receiving their first pulse at 45 °C, they survive a second high temperature pulse administered 1 day later. Thus, the acquisition of thermotolerance can prevent the development of thermosensitivity. The results suggest that some form of injury caused by the first challenge made the flies considerably more vulnerable to the second challenge. What type of injury is inflicted by the first pulse is unclear. Without the second challenge, the injury can apparently be repaired, but the problem arises if the insect is challenged a second time before it has fully recovered. The cause is not likely to involve an impairment of the heat shock protein response because thermosensitized flies are fully capable of synthesizing heat shock proteins.

Most mammalian cell lines develop thermotolerance when treated at temperatures below 42-43 °C but become thermosensitive when treated at temperatures above this range (Jung & Kolling 1980, Jung 1982, Nielsen et al. 1982, Dikomey et al. 1984). Chinese hamster ovarian cells exposed to 40 °C for 1-16 h develop thermotolerance to 43 °C, whereas a 15-90 min exposure to 43 °C induces thermosensitivity to 40 °C (Jung & Kolling 1980, Jung 1982). Survival curves for the thermosensitized mammalian cells are similar to the survival curve for thermosensitized flesh flies. In both cases, survival drops off quickly as duration of the second exposure increases. In

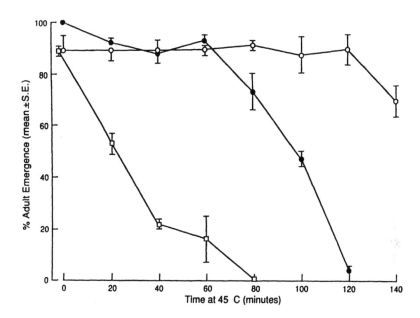

FIGURE 2.10 Thermosensitivity generated by two temporally separated exposures to high temperature. Flesh flies *Sarcophaga crassipalpis* reared at 25°C were exposed to 40°C for 240 min (open circles), 45°C for 60 min (open squares), or remained at 25°C (solid circles). After 24 h at 25°C the flies were exposed to 45°C for varying durations. Flies were treated as pharate adults, and survival was based on the percent survival until adult emergence (mean±SE; three replicates of 15 flies each for each time point). From Yocum & Denlinger (1993).

contrast, the survival curves generated for cells or flies exposed to a single high temperature have an initial broad shoulder and then decline (Fig.2.2). The broad shoulder can perhaps be interpreted as a period of accumulation of non-lethal lesions, but at a certain point, any additional lesions result in death. The absence of a shoulder, as seen in the thermosensitized flies (Fig. 2.10), could thus imply that the organism or cells have no capacity to tolerate a series of lesions that would otherwise be non-lethal.

The intriguing practical implication of thermosensitivity is that the pattern of administering a thermal stress has important consequences for the insect's survival. Two relatively modest pulses of high temperature may be just as effective in causing death as a single pulse of a higher temperature. And, from an economic perspective, this type of wounding may require less energy input than needed to administer a single pulse of a higher temperature.

Future Directions

Responses to high temperature are among the best studied aspects of insect biology. Yet, many fundamental and intriguing questions remain unanswered. In the field, it is not at all clear how certain species can survive, and indeed thrive, at temperatures above 50 and even 60°C. Surely, the tissues of such insects are replete with undiscovered mechanisms for circumventing the problems we normally associate with high temperature. Like thermophilic bacteria, such insects may very well have enzymatic properties and other attributes deserving study.

We still lack definitive proof of the nature of heat injury. What sites actually are the weak links leading to death? Do the overt manifestations of injury (e.g., failure of adult eclosion, failure of reproduction) share a common basis? An enormous body of literature documents the heat shock response, but the exact role of the heat shock proteins in the development of thermotolerance remains poorly understood. Why are there so many different heat shock proteins, and what roles do each play? Why would different tissues of the same organism or different developmental stages express different sets of proteins? We are far from understanding the full complexity of the heat shock response. And, it is also clear that thermotolerance involves more than just heat shock proteins. The correlations between thermotolerance and expression of the heat shock proteins are not perfect. Thermotolerance sometimes persists long after the heat shock proteins disappear. Yet, we know little about other metabolic adjustments that contribute to thermotolerance.

The fact that individual cells and tissues can respond directly and independently to high temperature implies that each cell has a mechanism for measuring temperature, a thermometer if you will, that regulates the cell's response. This, however, does not negate the possibility that the CNS and related endocrine glands also contribute to a coordinated response. The possibility that stress hormones may participate in an insect's high temperature response seems likely but remains largely uninvestigated.

Survival curves which plot survival against duration of high temperature exposure characteristically have a broad shoulder (little mortality initially), followed by a high rate of death. We still lack a good explanation for these two phases of the curve. What prevents mortality initially? Is there a repair mechanism that is, at first, highly effective in combating the injury? And, what happens at the transition point that marks the onset of high mortality? Do nonlethal lesions accumulate to the point where they finally become lethal? The mechanisms of themotolerance are clearly vulnerable to inactiviation.

Insects that have been thermosensitized are highly susceptible to thermal injury. The hallmark of such insects is their loss of thermal tolerance, but at this point, we have few insights into the nature of the injury that renders them susceptible to high temperature injury.

If thermotolerance conveys such an advantage to an insect, why do insects not simply maintain their bodies in a thermotolerant state all the time? The expense of doing so must be too great. Characteristically the thermotolerance mechanisms are turned on quickly and again promptly shut down once the challenge has passed. This implies a fascinating system of trade-offs, and an incompatibility of the thermotolerance mechanism and the activities that need to go on at normal temperatures. We know little about such trade-offs, yet they could provide interesting insight into the tolerance mechanisms and the evolution of thermotolerance. How readily do insect populations respond to temperature changes? Insects offer a great model for probing the potential impact of temperature changes, such as those involved in global warming. And, is the natural variation in thermotolerance great enough to suggest that high temperature manipulations used for insect control might result in the selection of highly thermotolerant populations of insect pests?

From the perspective of integrated pest management, a number of physiological responses to high temperature suggest potential for exploitation. Experimentally, we have shown that thermotolerance can be prevented. Application of a high temperature stress in a non-oxygenated atmosphere prevents the acquisition of thermotolerance. Likewise, an insect can be thermosensitized and thus be vulnerable to injury at relatively modest high temperatures. Such combination treatments (e.g., heat plus anoxia) or thermosensitization (e.g., two temporally separated treatments at moderate temperature conditions) are especially attractive because they can cause mortality with less energy input (and hence, less expense) than if heat, alone, is used. Evidence from human medicine suggests a wealth of chemical tools with potential to render an organism more vulnerable to thermal injury. Such tools for insects remain to be discovered.

References

Ahearn, G. A. 1970. Changes in hemolymph properties accompanying heat death in the desert tenebrionid beetle *Centrioptera muricata.* Comparative Biochem. Physiol. 33: 845-857.

Alexandrov, V. Y. 1977. *Cells, Molecules and Temperature: Conformational Flexibility of Macromolecules and Ecological Adaptation.* Springer-Velag, New York.

Anderson, J. F. & W. R. Horsfall. 1963. Thermal stress and anomalous development of mosquitoes (Diptera: Culicidae). I. Effect of constant temperature of dimorphism of adults of *Aedes stimulans*. J. Experimental Biol. 154: 67-107.

Anderson, R. L., R. G. Ahier & J. M. Littleton. 1983. Observations on the cellular effects of ethanol and hyperthermia *in vivo*. Radiation Res. 94: 318-325.

Ando, M., K. Katagiri, S. Yamamoto, S. Asanuma, M. Usuda, I. Kawahara & K. Wakamatsu. 1994. Effect of hyperthermia on glutathione peroxidase and lipid peroxidative damage in liver. J. Thermal Biol. 19: 177-185.

Arancia, G., P. Crateri Ttrovalusci, G. Mariutti & B. Mondovi. 1989. Ultrastructural changes induced by hyperthermia in Chinese hamster V79 fibroblasts. Int. J. Hyperthermia 5: 341-350.

Arrigo, A. -P. & J. Landry. 1994. Expression and function of the low-molecular-weight heat shock proteins, *In* R. I. Morimoto, A. Tissieres & C. Georgopoulos, eds. *The Biology of Heat Shock Proteins and Molecular Chaperones*. pp 335-373. Cold Spring Harbor Laboratory Press, New York.

Ashburner, M. & J. J. Bonner. 1979. The induction of gene activity in *Drosophila* by heat shock. Cell 17: 241-254.

Atkinson, B. G., T. Cunningham, R. L. Dean & M. Somerville. 1983. Comparison of the effects of heat shock and metal-ion stress on gene expression in cells undergoing myogenesis. Can. J. Biochem. Cell Biol. 61: 404-413.

Attfield, P. V. 1987. Trehalose accumulates in *Saccharomyces cerevisiae* during exposure to agents that induce heat shock response. FEBS Letters 225: 259-263.

Bell, C. H. 1983. Effect of high temperatures on larvae of *Ephestia elutella* (Lepidoptera: Pyralidae) in diapause. J. Stored Product Research 19: 153-157.

Berger, E. M. & M. P. Woodward. 1983. Small heat shock proteins in *Drosophila* may confer thermal tolerance. Experimental Cell. Res. 147: 437-442.

Bowler, K. 1963. A study of the factors involved in acclimatization to temperature and death at high temperatures in *Astacus pallipes*. II. Experiments at the tissue level. J. Cell Comparative Physiol. 62: 133-146.

_____. 1987. Cellular heat injury: Are membranes involved? *In* K. Bowler & B. J. Fuller, eds. *Temperature and Animal Cells*, pp 157-185. Soc. Experimental Biol. Symposium 41, Cambridge, England.

Bresnick, E., F. C. Dalman, E. R. Sanchez & W. B. Pratt. 1989. Evidence that the 90-kDa heat shock protein is necessary for the steroid binding conformation of the L cell glucocorticoid receptor. J. Biol. Chem. 264: 4992-4997.

Burton, V., H. K. Mitchell, P. Young & N. S. Petersen. 1988. Heat shock protection against cold stress of *Drosophila melanogaster*. Mol. Cell. Biol. 8: 3550-3552.

Calderwood, S. K., M. A. Stevenson & G. M. Hahn. 1988. Effects of heat on cell calcium and inositol lipid matabolism. Radiation Res. 113: 414-425.

Cavicchi, S., D. Guerra, V. La Torre & R. B. Huey. 1995. Chromosomal analysis of heat-shock tolerance in *Drosophila melanogaster* evolving at different temperatures in the laboratory. Evolution 49: 676-684.

Chen, C.-P., D. L. Denlinger & R. E. Lee. 1987. Cold shock injury and rapid cold-hardening in the flesh fly, *Sarcophaga crassipalpis*. Physiol. Zool. 60: 297-304.

Chen, C.-P., R. E. Lee & D. L. Denlinger. 1990. A comparison of the response of tropical and temperate flies (Diptera: Sarcohagidae) to cold and heat stress. J. Comparative Physiol. B 160: 543-547.

_____. 1991. Cold shock and heat shock: A comparison of the protection generated by brief pretreatment at less severe temperatures. Physiol. Entomol. 16: 19-26.

Coakley, W. T. 1987. Hyperthermia effects on the cytoskeleton and on cell morphology, *In* K. Bowler & B. J. Fuller, eds. *Temperature and Animal Cells.* pp 187-211, Soc. Experimental Biol. Symposium 41, Cambridge, England.

Cohen, A. & R. Patana. 1982. Ontogenetic and stress-related changes in hemolymph chemistry of beet armyworm. Comparative Biochem. Physiol. A71: 193-198.

Colaco, C., S. Sen, M. Thangavelu, S. Pinder & B. Roser. 1992. Extraordinary stability of enzymes dried in trehalose: Simplified molecular biology. Biotechnology 10: 1007-1011.

Cowley, J. M., K. D. Chadfield & R. T. Baker. 1992. Evaluation of dry heat as a postharvest disinfestation treatment for persimmons. New Zealand J. Crop and Hort. Sci. 20: 209-215.

Craig, E. A. 1985. The heat shock response. CRC Critical Review Biochem. 18: 239-280.

Craig, E. A., B. D. Gambill & R. J. Nelson. 1993. Heat shock proteins: molecular chaperones of protein biogenesis. Microbiol. Rev. 57: 402-414.

Darby, H. H. & E. M. Kapp. 1933. Observation on the thermal death points of *Anastrepha ludens* (Loew). USDA Tech. Bull 400.

Davenport, A. K. & P. D. Evans. 1984. Stress-induced changes in octopamine levels of insect haemolymph. Insect Biochem. 14: 135-143.

Denlinger, D. L., J. Giebultowicz & T. Adedokun. 1988. Insect diapause: dynamics of hormone sensitivity and vulnerability to environmental stress, *In* F. Sehnal, A. Zabza & D. L. Denlinger, eds. *Endocrinological Frontiers in Physiological Insect Ecology.* pp 243-262. Wroclaw Technical University Press, Wroclaw, Poland.

Denlinger, D. L., K. H. Joplin, C. -P. Chen & R. E. Lee. 1991. Cold shock and heat shock, *In* R. E. Lee & D. L. Denlinger, eds. *Insects at Low Temperature.* pp 131-148. Chapman and Hall, New York.

Dikomey, E., J. Eickhoff & H. Jung. 1984. Thermotolerance and thermosensitization in CHO and R1H cells: A comparative study. Int. J. Radiation Biol. 46: 181-192.

Downer, R. G. & V. L. Kallapur. 1981. Temperature-induced changes in lipid composition and transition temperature of flight muscle mitochondria of *Schistocerca gregaria*. J. Thermal Biol. 6: 189-194.

Dura, J. M. 1981. Stage dependent synthesis of heat shock induced proteins in early embryos of *Drosophila melanogaster*. Mol. Gen. Genet. 184: 381-385.

Evans, D. E. 1981. The influence of some biological and physical factors on the heat tolerance relationships for *Rhyzopertha dominica* (F.) and *Sitophilus oryzae* (L.)

(Coleoptera: Bostrychidae and Curculionidae). J. Stored Product Research 17: 65-72.

Feder, J. H., J. M. Rossi, J. Solomon, N. Solomon & S. Lindquist. 1992. The consequences of expressing hsp70 in *Drosophila* cells at normal temperatures. Genes Dev. 6: 1402-1413.

Fisher, G. A., R. L. Anderson & G. M. Hahn. 1986. Glucocorticoid-induced heat resistance in mammalian cells. J. Cell. Physiol. 128: 127-132.

Fittinghoff, C. M. & L. M. Riddiford. 1990. Heat sensitivity and protein synthesis during heat-shock in the tobacco hornworm, *Manduca sexta.* J. Comparative Physiol. B 160: 349-356.

Fraenkel. G. S. & D. L. Gunn. 1961. *The Orientation of Animals.* Dover, New York.

Friedlander, M. J., N. Kotchabhaki & C. L. Prosser. 1976. Effects of cold and heat on behavior and cerebellar function in goldfish. J. Comparative Physiol. A 112: 19-45.

Gladwell, R. T., K. Bowler & C. J. Duncan. 1975. Heat death in the crayfish *Austropotamobius pallipes* -ion movements and their effects on excitable tissues during heat death. J. Thermal Biol. 1: 79-94.

Gonen, M. 1977. Susceptibility of *Sitophilus granarius* and *S. oryzae* (Coleoptera: Curculionidae) to high temperature after exposure to supra-optimal temperature. Entomologia Experimentalis et Applicata 21: 243-248.

Grainger, J. N. R. 1973. A study of heat death in nerve-muscle preparation of *Rana temporaria.* Proc. Royal Irish Academy 73: 283-290.

Greenspan, R. J., J. A. Finn, Jr. & J. C. Hall. 1980. Acetylcholinesterase mutants in *Drosophila* and their effects on the structure and function of the central nervous system. J. Comparative Neurol. 189: 741-774.

Hadley, N. R. 1994. *Water Relations of Terrestial Arthropods.* Academic Press, San Diego.

Hadorn, E. 1955. Lethalfaktoren. pp 233-253. Georg Thieme Verlag, Stuttgart.

Hartke, A., S. Boucher, J.-M. Laplace, A. Benachour, P. Boutibonnes & Y. Auffray. 1995. UV-inducible proteins and UV-induced cross-protection against acid, ethanol, H_2O_2 or heat treatments in *Lactococcus lactis* subsp. *lactis.* Arch. Microbiol. 163: 329-336.

Heinrich, B. 1980. Mechanisms of body-temperature regulation in honeybees. *Apis mellifera* II. Regulation of thoracic temperature at high air temperatures. J. Experimental Biol. 85: 73-87.

_____. 1993. *The Hot-Blooded Insects.* Harvard University Press, Cambridge.

Heinrich, B. & G. A. Bartholomew. 1971. An analysis of pre-flight warm-up in the sphinx moth, *Manduca sexta.* J. Experimental Biol. 55: 223-239.

Heinrich, B. & T. M. Casey. 1978. Heat transfer in dragonflies: 'fliers' and 'perchers'. J. Experimental Biol. 74: 17-36.

Henle, K. J. 1981. Interaction of mono- and polyhydroxy alcohols with hyperthermia in CHO cells. Rad. Res. 88: 392-402.

Henle, K. J., T. P. Monson, W. A. Nagle & A. J. Moss. 1985. Heat protection by sugars and sugar analogues. Int. J. Hyperthermia 1: 371-382.

Henle, K. J., J. W. Peck & R. Higashikubo. 1983. Protection against heat-induced cell killing by polyols *in vitro.* Cancer Res. 43: 1624-1627.

Henle, K. J. & R. L. Warters. 1982. Heat protection by glycerol *in vitro.* Cancer Res. 42: 2171-2176.

Hochachka, P. W. & G. N. Somero. 1984. *Biochemical Adaptation.* Princeton University Press, Princeton.

Hoffmann, A. A. 1995. Acclimation: Increasing survival at a cost. Trends Evol. Ecol. 10: 1-2.

Hoffmann, A. A., H. Dahger, M. Hercus & D. Berrigan. 1997. Comparing different measures of heat resistance in selected lines of *Drosophila melanogaster.* J. Insect Physiol. 43: 393-405.

Hoffmann, G. E. & G. N. Somero. 1995. Evidence for protein damage at environmental temperatures: Seasonal changes in levels of ubiquitin conjugates and hsp70 in the intertidal mussel *Mytilus trossulus.* J. Experimental Biol. 198: 1509-1518.

Hoffman, A. A. & M. Watson. 1993. Geographical variation in the acclimation responses of *Drosophila* to temperature extremes. Amer. Naturalist 142: S93-S113.

Horsfall, W. R., J. F. Anderson & R. A. Brust. 1964. Thermal stress and anomalous development of mosquitoes (Diptera: Culicidae) III. *Aedes sierrensis.* Can. Entomologist 96: 1369-1372.

Hottiger, T., T. Boller & A. Wiemken. 1987. Rapid changes of heat and desiccation tolerance correlated with changes of trehalose content in *Saccharomyces cerevisiae* cells subjected to temperature shifts. FEBS Letters 220: 133-115.

_____. 1989. Correlation of trehalose content and heat resistance in yeast mutants altered in the RAS/adenylate cyclase pathway: Is trehalose a thermoprotectant? FEBS Letters 255: 431-434.

Huey, R. B. & J. G. Kingsolver. 1989. Evolution of thermal sensitivity of ectotherm performance. Trends Ecol. Evol. 4: 131-132.

_____. 1993. Evolution of resistance to high temperature in ectotherms. Amer. Naturalist 142: S21-S46.

Huey, R. B., L. Partridge & K. Fowler. 1991. Thermal sensitivity of *Drosophila melanogaster* responds rapidly to laboratory natural selection. Evolution 45: 751-756.

Hutchison, K., M. Czar, L. Scherrer & W. Pratt. 1992. Monovalent cation selectivity for the ATP-dependent association of the glucocorticoid receptor with hsp70 and hsp90. J. Biol. Chem. 267: 14047-14053.

Ishikawa, H. 1973. Comparative studies on the thermal stability of animal ribosomal RNAs. Comparative Biochem. Physiol. 46B: 217-277.

_____. 1975. Comparative studies on the thermal stability of animal ribosomal RNA-II. Sea-anemones (Coelenterata). Compar. Biochem. Physiol. 50B: 1-4.

Ivanovic, J.1991. Metabolic response to stressors, *In* J. Ivanovic & M. Jankovic-Hladni, eds. *Hormones and Metabolism in Insect Stress*. pp 27-67, CRC Press, Boca Raton, Florida.

Ivanovic, J. & M. Jankovic-Hladni, eds. 1991. *Hormones and Metabolism in Insect Stress*. CRC Press, Boca Raton, Florida.

Jaenicke, R. 1991. Protein stability and molecular adaptation to extreme conditions. Eur. J. Biochem. 202: 715-728.

Jang, E. B. 1986. Kinetics of thermal death in eggs and first instars of three species of fruit flies (Diptera: Tephritidae). J. Econ. Entomol. 79: 700-705.

_____. 1991. Thermal death kinetics and heat tolerance in early and late third instars of the Oriental fruit fly (Diptera: Tephritidae). J. Econ. Entomol. 84: 1298-1303.

Joplin, K. H. & K. L. Denlinger. 1990. Developmental and tissue specific control of the heat shock induced 70kDa related proteins in the flesh fly, *Sarcophaga crassipalpis*. J. Insect Physiol. 36: 239-249.

Joplin, K. H., G. D. Yocum & D. L. Denlinger. 1990. Cold shock elicits expression of heat shock proteins in the flesh fly, *Sarcophaga crassipalpis*. J. Insect Physiol. 36: 825-834.

Jung, H. 1982. Interaction of thermotolerance and thermosensitization induced in CHO cells by combined hyperthermic treatment at 40°C and 43°C. Radiation Res. 91: 433-466.

Jung, H. & H. Kolling. 1980. Induction of thermotolerance and thermosensitization in CHO cells by combined hyperthermic treatment at 40°C and 43°C. Eur. J. Cancer 16: 1523-1528.

Kim, S. H., S. S. Hong, A. A. Alfieri & J. H. Kim. 1988. Interaction of hyperthermia and pentamidine in HeLa S-3 cells. Radiation Res. 116: 320-326.

Kim, S. H., J. H. Kim & E. W. Hahn. 1978. Selective potentiation of hyperthermic killing of hypoxic cells by 5-thio-D-glucose. Cancer Res. 38: 2935-2938.

Kim, D. & Y. J. Lee. 1993. Effect of glycerol on protein aggregation quantitation of thermal aggregation of proteins from CHO cells and analysis of aggregated proteins. J. Thermal Biol. 18: 41-48.

Kingsolver, J. G. 1985. Thermoregulatory significance of wing melanization in *Pieris* butterflies (Lepidoptera: Pieridae): Physics, posture, and pattern. Oecologia 66: 546-553.

Koga, T. & K. Takumi. 1995. Comparison of cross-protection against some environmental stresses between cadmium-adapted and heat-adapted cells of *Vibrio parahaemolyticus*. J. Gen. Appl. Microbiol. 41: 263-268.

Koishi, M., N. Hosokawa, M. Sato, A. Nakai, K. Hirayoshi, M. Hiraoka, M. Abe & K. Nagata. 1992. Quercetin, an inhibitor of heat shock protein synthesis, inhibits the acquisition of thermotolerance in a human colon carcinoma cell line. Jpn. J. Cancer Res. 83: 1216-1222.

Krebs, R. A. & M. E. Feder. 1997. Natural variation in the expression of the heat-shock protein HSP70 in a population of *Drosophila melanogaster* and its correlation with tolerance of ecologically relevant thermal stress. Evolutions 51: 173-179.

Krebs, R. A. & M. E. Feder. 1998. HSP70 and larval thermotolerance in *Drosopihila melanogaster*: how much is enough and when is more too much? J. Insect Physiol. (in press).

Krebs, R. A. & V. Loeschcke. 1994. Costs and benefits of activation of the heat-shock response in *Drosophila melanogaster*. Functional Ecol. 8: 730-737.

Laszlo, A. 1992. The effects of hyperthermia on mammalian cell structure and function. Cell Prolif. 25: 59-87.

Lee, D. C. & D. Chapman. 1987. The effects of temperature on biological membranes and their models, *In* K. Bowler & B. J. Fuller, eds. *Temperature and Animal Cells*, pp 35-52. Soc. Experimental Biol. Symposium 41, Cambridge, England.

Lee, Y. J. & W. C. Dewey. 1988. Thermotolerance induced by heat, sodium arsenite, or puromycin: Its inhibition and differences between 43°C and 45°C. J. Cellular Physiol. 135: 397-406.

Lepock, J. R., K.-H. Cheng, H. Al-Qysi, I. O. Sim, C. J. Koch & J. Kruuv. 1987. Hyperthermia-induced inhibition of respiration and mitochondrial protein denaturation in CHL cells. Int. J. Hyperthermia 3: 123-132.

Leroi, A. M., A. F. Bennett & R. E. Lenski. 1994. Temperature acclimation and competitive fitness: An experimental test of the beneficial acclimation assumption. Proc. Natl. Acad. Sci., USA 91: 1917-1921.

Lesham, Y. Y. & P. J. C. Kuiper. 1996. Is there a GAS (general adaptation syndrome) response to various types of environmental stress? Biologia Plantarum 38: 1-18.

Levins, R. 1969. Thermal acclimation and heat resistance in *Drosophila* species. Amer. Naturalist 103: 483-499.

Lindquist, S. 1986. The heat-shock response. Annual Review Biochem. 55: 1151-1191.

Lindquist, S. & E. A. Craig. 1988. The heat-shock proteins. Annual Review Genetics 22: 631-677.

Malcolm, N. L. 1969. Molecular determinants of obligate psychrophily. Nature 221: 1031-1033.

Mellanby, K. 1954. Acclimatization and the thermal death in insects. Nature 1973: 582.

Mikkelsen, R. B., T. Stedman, R. Schmidt-Ulrich & P.-S. Lin. 1991. Effects of tamoxifen and protein kinase C inhibitors on hyperthermic cell killing. Int. J. Oncol. Biol. Phys. 20: 1039-1045.

Milkman, R. 1962. Temperature effects on day old *Drosophila* pupae. J. Gen. Physiol. 45: 777-799.

_____. 1963. On the mechanism of some temperature effects on *Drosophila*. J. Gen. Physiol. 46: 1151-1170.

Milkman, R. & B. Hille. 1966. Analysis of some temperature effects on *Drosophila* pupae. Biol. Bull. 13: 331-345.

Mitchell, H. K. 1966. Phenol oxidases and *Drosophila* development. J. Insect Physiol. 12: 755-765.

Mitchell, H. K. & L. S. Lipps. 1978. Heat shock and phenocopy induction in *Drosophila*. Cell 15: 907-918.

Mitchell, H. K., G. Moller, N. S. Petersen & L. Lipps-Sarmiento. 1979. Specific protection from phenocopy induction by heat shock. Dev. Genet. 1: 181-192.

Morimoto, R. I., A. Tissieres & C. Georgopoulos, eds. 1994. *The Biology of Heat Shock Proteins and Molecular Chaperones.* Cold Spring Harbor Laboratory Press, New York.

Moss, J. I. & E. B. Jang. 1991. Effects of age and metabolic stress on heat tolerance of Mediterranean fruit fly (Diptera: Tephritidae) eggs. J. Econ. Entomol. 84: 537-541.

Mutero, A., J.-M. Bride, M. Pralavorio & D. Fournier. 1994. *Drosophila melanogaster* acetylcholinesterase: Identification and expression of two mutations responsible for cold- and heat-sensitive phenotypes. Mol. Gen. Genet. 243: 699-705.

Nielsen, O. S., K. J. Henle & J. Overgaard. 1982. Arrhenius analysis of survival curves from thermotolerant and step-down heated L1A2 cells *in vitro*. Radiation Res. 91: 468-482.

O'Kasha, A. Y. K. 1968a. Effect of sub-lethal high temperature on insect, *Rhodnius prolixus* (Stål). I. Induction of delayed moulting and defects. J. Experimental Biol. 48: 455-463.

———. 1968b. Effect of sub-lethal high temperature on insect *Rhodnius prolixus* (Stål). II. Mechanism of cessation and delay of moulting. J. Experimental Biol. 48: 464-473.

———. 1968c. Effect of sub-lethal high temperature on insect *Rhodnius prolixus* (Stål). III. Metabolic changes and their bearing on the cessation and delay of moulting. J. Experimental Biol. 48: 475-486.

Othsuka, K., Y.-C. Liu & T. Kaneda. 1993. Cytoskeletal thermotolerance in NRK cells. Int. J. Hyperthermia. 9: 115-124.

Parsell, D. A. & S. Lindquist. 1993. The function of heat-shock proteins in stress tolerance: Degradation and reactivation of damaged proteins. Annual Review Genetics 27: 437-496.

———. 1994. Heat shock proteins and stress tolerance, *In* R. Morimoto, A. Tissieres & C. Georgopoulos, eds. *The Biology of Heat Shock Proteins and Molecular Chaperones,* pp 457-494, Cold Spring Harbor Laboratory Press, New York.

Parsell, D. A. & R. T. Sauer. 1989. Induction of a heat shock-like response by unfolded protein in *Escherichia coli:* Dependence on protein level not protein degradation. Genes Dev. 3: 1226-1232.

Patil, N. S., K. S. Lole & D. N. Deobagkar. 1996. Adaptive larval thermotolerance and induced cross-tolerance to propoxur insecticide in mosquitoes *Anophleles stephensi* and *Aedes aegypti*. Med. Vet. Entomol. 10: 277-282.

Petersen, N. S. & H. K. Mitchell. 1985. Heat-shock proteins, *In* G. A. Kerkut & L. I. Gilbert, eds. *Comprehensive Insect Physiology, Biochemistry, and Pharmacology.* vol. 10, pp 347-365, Pergamon, New York.

Prange, H. D. 1996. Evaporative cooling in insects. J. Insect Physiol. 42: 493-499.

Prosser, C. L. 1986. *Adaptational Biology: Molecules to Organisms.* Wiley, New York.

Rauschenbach, I. Y. 1991. Changes in juvenile hormone and ecdysteroid content during insect development under heat stress, *In* J. Ivanovic & M. Jankovic-Hladni, eds. *Hormones and Metabolism in Insect Stress,* pp 115-148, CRC Press, Boca Raton, Florida.

Rauschenbach, I. Y., L. I. Serova, I. S. Timochina, N. A. Chentsova & L. V. Schumnaja. 1993. Analysis of differences in dopamine content between two lines of *Drosophila virilis* in response to heat stress. J. Insect Physiol. 39: 761-767.

Reiber, C. L. & G. F. Birchard. 1993. Effect of temperature on metabolism and hemolymph pH in the crab *Stoloczia abbotti.* J. Thermal Biol. 18: 49-52.

Reiter, T. & S. Penman. 1983. "Prompt" heat shock proteins: Translationally regulated synthesis of new proteins associated with the nuclear matrix-intermediate filaments as an early response to heat shock. Proc. Natl. Acad. Sci., USA 80: 4737-4741.

Robertson, R. M., H. Xu, K. L. Shoemaker & K. Dawson-Scully. 1995. Exposure to heat shock affects thermosensitivity of the locust flight system. J. Neurobiol. 29: 367-383.

Roti Roti, J. L. 1982. Heat-induced cell death and radiosensitization: Molecular mechanisms, *In* L. A. Dethlefsen & W. C. Dewey, eds. *Proceedings of the Third International Symposium: Cancer Therapy by Hyperthermia, Drugs and Radiation.* pp 3-10, National Cancer Institute Monographs 61.

Schmitz, H. & L. T. Wasserthal. 1993. Antennal thermoreceptors and wing thermosensitivity of heliotherm butterflies: Their possible role in thermoregulatory behavior. J. Insect Physiol. 39: 1007-1019.

Sharp, J. L. 1986. Hot-water treatment for control of *Anastrepha suspensa* (Diptera: Tephritidae) in mangoes. J. Econ. Entomol. 79: 706-708.

Sharp, J. L. & H. Picho-Martinez. 1990. Hot-water quarantine treatment to control fruit flies in mangoes imported into United States from Peru. J. Econ. Entomol. 83: 1940-1943.

Shukla, R. M., G. Chand & M. L. Saini. 1989. Effect of malathion resistance on tolerance to various environmental stresses in rust-red flour beetle (*Tribolium castaneum*). Indian J. Agric. Sciences 59: 778-780.

Smith, D. F. & D. O. Toft. 1993. Steroid receptors and their associated proteins. Mol. Endocrinol. 7: 4-11.

Soderstrom, E. L., D. G. Brand & B. Mackey. 1992. High temperature combined with carbon dioxide enriched or reduced oxygen atmospheres for control of *Tribolium castaneum* (Herbst) (Coleoptera: Tenebrionidae). J. Stored Product Research 28: 235-238.

Solomon, J. M., J. M. Rossi, K. Golic, T. McGarry & S. Lindquist. 1991. Changes in Hsp70 alter thermotolerance and heat-shock regulation in *Drosophila.* New Biologist 3: 1106-1120.

Somero, G. N. 1995. Proteins and temperature. Annual Review Physiol. 57: 43-68.

Streffer, C. 1985. Metabolic changes during and after hyperthermia. Int. J. Hyperthermia 1: 305-319.

Ungermann, C., W. Neupert & D. M. Cyr. 1994. The role of hsp70 in conferring unidirectionality on protein translocation into mitochondria. Science 266: 1250-1253.

Ushakov, B. 1964. Thermostability of cells and proteins of poikilotherms and its significance in speciation. Physiol. Review 44: 518-559.

Uvarov, B. P. 1931. Insects and climate. Trans. Entomol. Soc. London 79: 1-247.

Velazquez, J. M., S. Sonoda, G. Bugaisky & S. Lindquist. 1983. Is the major *Drosophila* heat shock protein present in cells that have not been heat shocked? J. Cell Biol. 96: 286-290.

Vidair, C. A., S. J. Doxesya & W. C. Dewey. 1993. Heat shock alters centrosome organization leading to mitotic dysfunction and cell death. J. Cell. Physiol. 154: 443-455.

Walter, M. F., N. S. Petersen & H. Biessmann. 1990. Heat shock causes the collapse of the intermediate filament cytoskeleton in *Drosophila* embryos. Dev. Genet. 11: 270-279.

Warter, R. L. & L. M. Brizgys. 1987. Apurinic site induction in the DNA of cells heated at hyperthermic temperature. J. Cell Physiol. 133: 144-150.

Wehner, R. A., C. Marsh & S. Wehner. 1992. Desert ants on a thermal tightrope. Nature 357: 586-587.

Welte, M. A., J. M. Tetrault, R. P. Dellavalle & S. Lindquist. 1993. A new method for manipulating transgenes: Engineering heat tolerance in a complex multicellular organism. Current Biol. 3: 842-853.

White, R. L. 1983. Effects of acute temperature change and acclimation temperature on neuromuscular function and lethality in crayfish. Physiol. Zool. 56: 174-194.

Whiting, D. C., S. P. Foster & J. H. Maindonald. 1991. Effects of oxygen, carbon dioxide, and temperature on the mortality responses of *Epiphyas postvittana* (Lepidoptera: Tortricidae). J. Econ. Entomol. 84: 1544-1549.

Wharton, G. B. 1985. Water balance of insects, *In* G. A. Kerkut & L. I. Gilbert, eds. *Comprehensive Insect Physiology, Biochemistry, and Pharmacology*, Vol. 14, pp 565-601. Pergamon, Oxford.

Wigglesworth, V. B. 1955. High temperature and arrested growth in *Rhodnius:* Quantitative requirements for ecdysone. J. Experimental Biol. 32: 649-663.

Willmer, P. G. 1982a. Microclimate and the environmental physiology of insects. Advances Insect Physiol. 16: 1-57.

_____. 1982b. Thermoregulatory mechanisms in *Sarcophaga*. Oecologia 53: 382-385.

Wolfe, G. R., D. L. Hendrix & M. E. Salvucci. 1998. A thermoprotective role for sorbitol in the silverleaf whitefly, *Bemisia argentifloii*. J. Insect Physiol. (in press).

Yi, P. N., C. S. Chang, M. Tallen, W. Bayer & S. Ball. 1983. Hyperthermia-induced intracellular ionic levels changes in tumor cells. Radiation Res. 93: 534-544.

Yocum, G. D. & D. L. Denlinger. 1992. Prolonged thermotolerance in the flesh fly, *Sarcophaga crassipalpis,* does not require continuous expression or persistence of the 72 kDa heat-shock protein. J. Insect Physiol. 38: 603-609.

_____. 1993. Induction and decay of thermosensitivity in the flesh fly, *Sarcophaga crassipalpis.* J. Comparative Physiol. B 163: 113-117.

_____. 1994. Anoxia blocks thermotolerance and the induction of rapid cold hardening in the flesh fly, *Sarcophaga crassipalpis.* Physiol. Entomol. 19: 152-158.

Yocum, G. D., J. Žďárek, K. H. Joplin, R. E. Lee, D. C. Smith, K. D. Manter & D. L. Denlinger. 1994. Alteration of the eclosion rhythm and eclosion behavior in the flesh fly, *Sarcophaga crassipalpis,* by low and high temperature stress. J. Insect Physiol. 40: 13-21.

Yoder, J. A., G. J. Blomquist & D. L. Denlinger. 1995. Hydrocarbon profiles from puparia of diapausing and non diapausing flesh flies (*Sarcophaga crassipalpis*) reflect quantitative rather than qualitative differences. Arch. Insect Biochem. Physiol. 28: 377-385.

Yoder, J. A. & D. L. Denlinger. 1991. Water balance in flesh fly pupae and water vapor absorption associated with diapause. J. Experimental Biol. 157: 273-286.

Yoder, J. A., D. L. Denlinger, M. W. Dennis & P. E. Kolattukudy. 1992. Enhancement of diapausing flesh fly paparia with additional hydrocarbons and evidence for alkane biosynthesis by a decarbonylation mechanism. Insect Biochem. Molec. Biol. 22: 237-243.

Yokoyama, V. Y., G. T. Miller & R. V. Dowell. 1991. Response of codling moth (Lepidoptera: Tortricidae) to high temperature, a potential treatment for exported commodities. J. Econ. Entomol. 84: 528-531.

3

Physiology of Cold Sensitivity

David L. Denlinger and Richard E. Lee, Jr.

Low temperatures pose a different set of challenges than those observed at high temperatures, albeit challenges every bit as formidable. While insects at high temperature are constantly threatened by high rates of water loss, at sub-zero temperatures insects are confronted with the obvious challenge of potential ice formation. To a small-bodied poikilotherm composed of roughly 70% water, management of body water becomes a critical issue at low temperatures. How can freezing be avoided or how can the body survive in a frozen state? And, numerous additional challenges to cell integrity and tissue function become evident as body temperature is lowered, even at temperatures well above 0 °C.

Defense against low temperature injury is evident at several levels. For a few insects, such as the monarch butterfly, the low temperatures of winter in North America are simply avoided by migration to a more moderate clime in the mountains of subtropical Mexico or southern California. But, for most insects, such an escape is not an option. Yet, the first line of defense, even for insects remaining in cold regions during the winter, is a behavioral response, a response that directs the overwintering insect to a thermally-buffered microenvironment. In addition to the selection of a thermally-favorable environment, insects can invoke an impressive array of physiologic mechanisms to prevent injury at low temperatures.

Insects can also exploit low temperatures for their own benefit. Bumble bees carrying a heavy load of conopid parasitoids stay away from the colony on cool nights and expose themselves to low temperatures, thus retarding development of the troublesome parasitoids and reducing the chances of successful parasitoid development (Müller & Schmid-Hempel 1993). And, on a more regular basis, the cold temperatures prevailing in winter likely enable many insects to escape pathogens that escalate in abundance during the favorable seasons of the year.

In this chapter we offer a brief overview of the injury inflicted by low temperature and the mechanisms used by insects to circumvent such injury. Protective mechanisms that might be disarmed to render the insect more vulnerable to low temperature injury are of particular interest in developing new strategies of insect pest management. Many aspects of low temperature responses were discussed previously in Lee & Denlinger (1991) and Leather et al. (1993). Diapause, a form of developmental arrest in common use by many overwintering insects, has been reviewed in considerable detail (Saunders 1982, Denlinger 1985, Tauber et al. 1986, Danks 1987, Zaslavski 1988).

Supercooling and Ice Nucleation

To understand the fundamental strategies of insects that overwinter at sub-zero temperatures it is necessary to consider the nature of supercooling and ice nucleation. These concepts have been treated extensively elsewhere (Angell 1982, Mazur 1984, Franks 1985, 1987, Karow 1991, Lee et al. 1991, 1993a, Vali 1995).

As an insect is cooled to sub-zero temperatures ice does not form at $0°C$, indeed it cannot form until the temperature falls below the melting point of the insect's body fluids (Fig. 3.1). For insects that have high concentrations of low-molecular-mass cryoprotectants the melting point of the blood may be colligatively depressed by many degrees. The beetle *Pytho deplanatus,* a species that dehydrates extensively during the winter, has a melting point of $-20°C$ (Ring 1982).

Insects only begin to supercool when they are cooled to temperatures below their melting point (Fig. 3.1). Small volumes of water supercool more readily than larger ones (Angell 1982, Vali 1995). Consequently, the small size of insects has allowed them to exploit this physical characteristic of supercooling in their overwintering strategies, whereas larger ectotherms such as amphibians and reptiles cannot (Costanzo & Lee 1995, 1996). The limit of supercooling, termed the supercooling point or temperature of crystallization, is reached when ice begins to form within the body fluids. This limit is easily detected by the appearance of the exotherm caused by the release of the heat of crystallization as body water freezes.

The supercooling point is generally regulated by the presence of endogenous, non-water ice nucleating agents (Lee 1991, Lee et al. 1996).

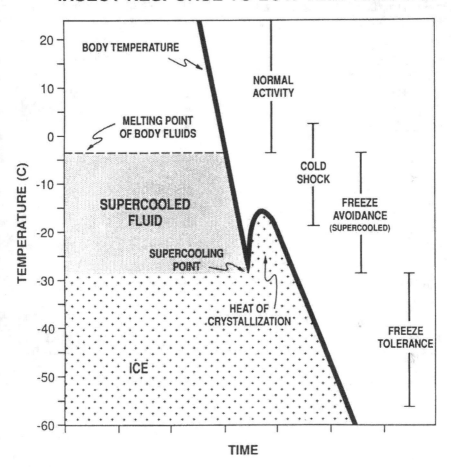

FIGURE 3.1 Responses of insects cooled to low temperature. The bold line indicates insect body temperature in relation to the melting point of body fluids, the supercooling point and the temperature at which internal ice forms. The right side of the figure indicates the general temperature ranges for different categories of insect response to low temperature. From Lee (1989).

These agents function as catalysts to promote ice nucleation at higher temperatures than would occur in their absence. Some freezing tolerant species synthesize proteins or lipoproteins that induce ice nucleation in the range of -6 to -9°C (Zachariassen 1992, Duman et al. 1995). The induction of ice nucleation at high sub-zero temperatures promotes survival of freezing by slowing the rate of ice formation compared to that which would occur if the insect supercooled extensively before freezing began. Mugnano et al. (1996) recently described a new class of insect nucleators in the form of endogenous crystals of calcium phosphate within the Malpighian tubules of the freeze tolerant fly larvae, *Eurosta solidaginis*. As discussed extensively in Chapter 4 recent reports of ice nucleating microorganisms isolated from insects identify yet another category of endogenous ice nucleators.

Classification of Insect Cold Hardiness

Investigations of cold tolerance typically place a given insect into one of two categories: the relatively few species that are freezing tolerant survive extensive internal ice formation, while freezing susceptible or intolerant insects succumb when their hemolymph freezes (Fig. 3.1). While these categories are useful in general discussions of insect cold-hardiness, they are simplistic when trying to describe the wide diversity of insect responses to low temperature (Baust & Rojas 1985, Lee 1991, Turnock & Bodnaryk 1991, Bale 1993). For example, some insects do not survive exposure to temperatures above 0°C. Others that supercool extensively and do not freeze until temperatures drop below -20°C may die at temperatures significantly above their supercooling points. Non-diapausing pupae of the flesh fly *Sarcophaga crassipalpis* supercool to -23°C but do not survive exposure to -17°C for even 20 min (Lee & Denlinger 1985). In contrast, diapausing pupae readily survive chilling at all temperatures above the supercooling point. Chilling-induced injuries may also be expressed as a consequence of long-term exposure (weeks or months) to low temperature. Turnock et al. (1985) reported reduced survival to eclosion in *Delia radicum*, a species with a supercooling point of -23°C, when it was held continuously at -10.2°C for 80 days.

Determining the type and level of cold tolerance for a given insect empirically is difficult for a number of reasons (Zachariassen 1985, Baust & Rojas 1985, Bale 1987, Block 1991, Lee 1991). Biological factors that must be considered include the developmental stage, diapause status and age, as well as the acclimation status of the species. Survival depends on both the exposure temperature and the duration of exposure. Spontaneous ice nucleation in the supercooled body fluids of an insect has a stochastic component (Salt 1961). For example, if a freeze intolerant insect is held for 1 h at $-20°C$ it might remain unfrozen and survive, however if the duration of exposure is extended to 24 h ice nucleation may occur and death results. To assess freezing tolerance the duration of exposure should be sufficient to allow an equilibrium level of internal ice formation to be attained; in the freeze tolerant gall fly *E. solidaginis* 2-3 days at $-23°C$ are required to reach this level (Lee & Lewis 1985). Cooling and warming rates have also been shown to critically influence survival (Miller 1978). Lastly, the criteria for survival following low temperature exposure must be biologically meaningful; the ability of an insect to wiggle or walk does not mean that it can survive to the next developmental stage, and that does not mean that the individual can ultimately reproduce and leave viable offspring - the best criterion for assessing cold tolerance.

Developmental Alterations Caused by Low Temperature

Mortality and failure to reproduce are not the only consequences of low temperature exposure. Low temperature also influences several aspects of development, including adult size, the number of larval instars, tissue morphogenesis, and sex ratio (Sehnal 1991).

As an insect larva grows, it normally progresses through a fixed number of instars. During its final larval instar, the larva attains a critical size that sets in motion the endocrine events that trigger metamorphosis (Nijhout & Williams 1974). The critical size may be reached very early in the final larval instar [e.g., *Sarcophaga bullata* (Žďárek & Sláma 1972)], or midway through the instar [e.g., *Glossina morsitans* (Denlinger & Žďárek 1991)]. Once the critical size has been reached, the larva is competent to initiate metamorphosis. For an insect such as *S. bullata*, which reaches its critical size early in the final larval instar, feeding can be halted long before the larva reaches its maximum size and it can still successfully proceed with metamorphosis. But, for larvae of *G. morsitans*, the critical size and maximum size are nearly the same: larvae removed from the food source prematurely fail to initiate metamorphosis. Since the final larval instar is frequently the longest instar and the period in which the most food is consumed, it is this instar that exerts a disproportionate influence on adult size. Adult size can thus be most readily influenced in species that attain their critical size early in the final instar.

Low temperatures that intercede after the critical size is attained may interfere with feeding processes and thus prevent the insect from reaching its maximum size, but the insect may still be able to initiate metamorphosis. Such a scenario results in the production of a small adult. For the carpet beetle, *Attagenus megatoma,* rearing at 20°C yields an adult that is only half the size of adults reared at 30-35°C (Baker 1983). Less time and energy are required to produce a small adult, and this may indeed be the basis for the small size characteristic of insects from alpine and polar environments (Danks 1981, Sømme & Block 1991). But, lowering the rearing temperature does not always result in smaller individuals. Within the range of 16-25°C, females of *D. melanogaster* are larger when reared at the lower end of the range (David et al. 1983), and in this case, larger size implies higher fecundity (Robertson 1957).

Low temperature may also indirectly influence adult size by influencing the number of larval instars. Though this number is rigidly fixed in most species, the number of instars in some species can be altered in response to low temperature or other environmental stresses. The number of instars can either be decreased or increased by low temperature. While the moth *Ephestia kuhniella* reared at 25°C normally has 5 larval instars, it pupates at the end of the fourth larval instar at 18°C (Gierke 1932). In the wax moth, *Galleria mellonella,* a cold shock (0°C for 30 min) at the beginning of what is normally the final larval instar prompts the larva to molt into an additional larval instar rather than pupate (Cymborowski & Bogus 1976). A decrease in the number of instars results in smaller adults, while an increase in the number of instars usually produces larger adults.

The cold shock that induces supernumerary molts in *G. mellonella* somehow alters the response of the regulatory centers within the brain. As a consequence, allatotropin is released at the wrong time (Cymborowski 1988), resulting in an elevated juvenile hormone titer (Sehnal & Rembold 1985), thus causing the subsequent molt to be a larval-larval molt rather than pupation.

Phenocopy defects, like those observed at high temperature (see Chapter 2), can also be elicited by low temperature, as demonstrated by the classic studies of Villee (1943, 1945) on temperature-sensitive homeotic mutants of *D. melanogaster.* In aristapedia (antennae are transformed into legs), low temperature rearing (15°C for several days after oviposition) shifts the direction of antennal development toward a tarsus, but formation of the normal antennal appendage (the arista) is favored at a higher temperature (29°C). In the homologous mutant, proboscipedia, exposure to the same low temperature regime causes the labial palps to be replaced with aristae, the antennal-like appendages. Interestingly, at higher temperatures (29°C) the labial palps are replaced with a tarsal-like appendage. The period during which the fly is susceptible to the effect of low temperature extends over several days and is most pronounced if the low temperature treatment is begun 5 days after oviposition. Thus, a much longer period of low temperature exposure is required to elicit phenocopy defects than is needed to elicit phenocopy defects at high temperature: while days of exposure are required at low temperature, only minutes are required at high temperature.

Different types of developmental defects can be elicited by low temperature at different developmental stages. Eggs of the chrysomelid beetle *Atrachya menetriesi* exposed to low temperatures divide into multiple embryos (Miya & Kobayashi 1971), a condition that is lethal. Low temperature during postembryonic development may cause the production of individuals with a mixture of larval and adult features (Sehnal 1991). The yellow mealworm, *Tenebrio molitor,* is quite vulnerable to cold-induced developmental aberrations (Lengerken 1932, Stellwaag-Kittler 1954). Cold treatment of last instar larvae can cause a molt that will produce a larva-like individual, but one that possesses rudimentary pupal-like eyes and appendages. Similar effect can be achieved by administration of juvenile hormone to final instar larvae (Sehnal & Schneiderman 1973), thus suggesting that the cold treatments elicit this developmental response by boosting the juvenile hormone titer at a time when the hormone should be absent or present only at low levels.

As noted with high temperature (Chapter 2), low temperatures frequently distort sex ratios (Lauge 1985, Wrensch 1993). Males of the psychid moth *Talaeporia tubulosa* are produced by eggs containing two sex chromosomes (XX), and females develop from eggs having only a single sex chromosome (XO). In the optimal temperature range females produce a nearly equal proportion of oocytes with and without the X chromosome, but when the female is reared at 3-5°C, the X chromosome is displaced to the polar body, and consequently most of the resulting progeny are females (Seiler 1920). In many Hymenoptera, fertilization is controlled by the female, and eggs that are not fertilized develop into females. In the chalcid *Ooencyrtus* low temperature during development favors the production of nonfertilized (female) eggs (Wilson & Woolcock 1960). A shift toward production of a higher proportion of males is also well documented in response to low temperature. A distinct, but slight, increase in male production in response to low temperature was noted for the citrus red mite, *Panonychus citri* (Munger 1963) and the icheumonid parasitoid *Campoletis perdistinctus* (Hoelscher & Vinson 1971).

In the ant *Formica rufa* the spermatheca fails to release sperm at temperatures below 19.5°C, thus only unfertilized eggs, in this case males, are produced (Gösswald & Bier 1955).

A shift in the autumn from parthenogenesis to sexual reproduction is common in thrips, aphids, cynipids, and many species of mites. Low temperature, in association with short daylength, frequently provides the cue triggering this shift to production of both males and females (Hardie & Lees 1985). In the autumn aphids also shift from apterous development to the formation of alates, a change that is again promoted by low temperature acting in concert with short daylength.

Diapause and Cold Tolerance

Most insects in the temperate zone are subjected to the lowest temperatures when they are in an overwintering diapause. The suppression of metabolism and purging of the gut (elimination of ice nucleators in the gut) are among characteristic features of diapause that can contribute to cold tolerance, yet diapause and cold hardiness are not consistently linked (Denlinger 1991). Diapause, by itself, does not necessarily imply cold hardiness, nor does cold hardiness imply that the insect is in diapause.

Diapause is not restricted to insects from temperate and polar regions, and it is not always limited to the winter season. Diapause is well documented in the tropics (Denlinger 1986) and can be expressed during the summer in temperate zones (Masaki 1980). In these situations diapause occurs in the apparent absence of cold hardening (Fig. 3.2). Metabolic suppression may result in altered carbohydrate metabolism in such cases (Pullin 1996), a feature often associated with increased cold hardening, but thus far no evidence is available demonstrating enhanced cold hardiness associated with either tropical or summer diapause.

At the opposite extreme, cold hardiness can readily be demonstrated in the absence of diapause. Examples include the development of cold hardiness in species that lack the capacity for diapause [e.g. *Tenebrio molitor* (Patterson & Duman 1978], cold hardening in nondiapausing stages of insects that do enter diapause [e.g. cold hardening in adults of the flesh fly *Sarcophaga bullata*, a species that diapauses as a pupa (Chen et al. 1987b)], and rapid cold hardening, the hardening response that can occur at any developmental stage within a few minutes of exposure to an intermediately low temperature (Chen et al. 1987a, Lee et al. 1987).

DIAPAUSE   COLD-HARDINESS

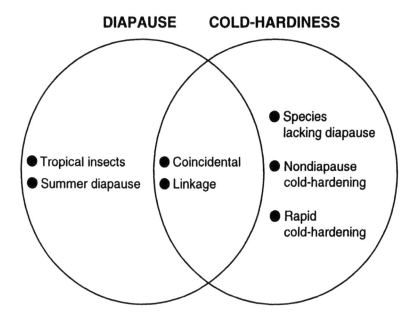

FIGURE 3.2 Relationship between diapause and cold hardiness. The two events may be expressed independently or in association with each other. When associated, the relationship may be coincidental or linked. Adapted from Denlinger (1991).

But, quite frequently diapause and cold hardiness are associated. This association can have two forms: diapause and cold hardiness may be only coincidentally associated or cold hardiness may actually be a component of the diapause program. The relationship is considered to be coincidental if separate environmental cues regulate diapause and cold hardiness. This is the case, for example, in the European corn borer, *Ostrinia nubilalis* (Hanec & Beck 1960): the larva enters diapause in response to short daylength but it becomes cold hardy only after it is exposed to low temperature. Separate environmental cues thus dictate these two events, and a corn borer can be in diapause without being cold hardy. For corn borers in the field, however, the expression of diapause and cold hardiness normally closely coincide.

By contrast, a firm linkage between diapause and cold hardiness is exemplified in the flesh flies *S. bullata* and *S. crassipalpis* (Adedokun & Denlinger 1984, Lee & Denlinger 1985). In these flies cold hardiness is a component of the diapause program. Flies that enter pupal diapause are already much more cold hardy than nondiapausing pupae. Entry into diapause is consistently linked to enhanced cold hardiness. A separate set of environmental cues is not needed to initiate the cold hardening process. Of course, low temperature may further enhance the cold hardening, but even without exposure to low temperature, the pupae are cold hardened.

Whether the cold hardiness is associated with diapause coincidentally or is linked to the diapause program, it is during diapause that most insects exhibit the greatest cold hardiness. For example, in *S. crassipalpis*, only diapausing pupae can survive at temperatures approaching its supercooling point (-23°C), and they can do so for many days (Lee & Denlinger 1985). Though the supercooling point is equally low in nondiapausing pupae, such pupae are readily killed following exposure to -10°C for less than an hour. In addition to the biochemical adaptations that contribute to cold hardiness during diapause, many diapausing species take refuge in thermally-buffered sites during diapause, and quite frequently they prepare special cocoons, hibernacula or other structures in which to overwinter (Danks 1991). This combination of developmental arrest, enhanced cold hardiness, selection and/or construction of a protected site results in an increased challenge when targeting diapausing individuals for thermal wounding.

Variation in Tolerance

A few classic examples illustrate the huge variation in cold tolerance that is evident among different species of insects and other arthropods. Certain species not only survive but remain active at temperatures near 0°C or lower. Snow fleas (Collembola) can be seen freely hopping over the surfaces of glaciers and snow fields at high altitudes. Likewise, grylloblattids are most active at temperatures near 0°C and will succumb when temperatures exceed 12°C (Morrissey & Edwards 1979). An antarctic mite *Nanorchestes antarcticus* remains active down to -11°C (Sømme & Block 1991), and a midge living in glacial pools in the mountains of Himalaya remains active at temperatures as low as -16°C (Kohshima 1984). Winter active moths (several species of noctuids and geometrids) continue to fly even when air temperatures drop as low as 0 to 10°C. Though the thoracic temperature of the moths during flight (30-35°C) is similar to flight temperature of other moths, the extraordinary feature of the winter moths is their ability to initiate the shivering needed for preflight warm-up at temperatures as low as 0°C (Heinrich 1987).

Variation in cold tolerance within a single population is evident from the success of genetic selection experiments. Tucic (1979) succeeded in selecting for greater cold tolerance in *Drosophila melanogaster*. By selecting for increased cold tolerance in one particular stage, he was able to increase cold hardiness in other stages as well, but the effect was diminished in stages more distant from the selected stage. In experiments by Chen & Walker (1994), separate lines of *D. melanogaster* were selected for greater tolerance against cold shock injury and long-term chilling injury. The cold-shocked line increased tolerance to cold shock, and the line selected for tolerance to long-term chill injury increased tolerance to long-term chilling injury. But interestingly, the increased tolerance to cold shock injury did not result in increased tolerance to long-term chilling injury, thus suggesting that these two forms of cold tolerance rely on distinct mechanisms.

Geographic variation is also evident. Though tropical species of flesh flies (Chen et al. 1990) and *Drosophila* (Hoffmann & Watson 1993) have some capacity for acclimating to low temperatures, the tropical species tend to be less cold tolerant than their temperate zone relatives, and even within the temperate region, populations at lower latitudes are less cold tolerant than those from higher latitudes [e.g., *Eurosta solidaginis* (Baust & Lee 1981, Lee et al. 1995)]. Ample evidence suggests a genetic basis for such differences in cold tolerance. A cold hardy species, *D. lutsescens*, crossed with a closely related species that is less cold hardy, *D. takahashii*, yields progeny with an intermediate level of cold hardiness (Kimura 1982). Variation of cold hardiness that is inherent in a natural population can provide the grist for selecting strains of insects with increased cold tolerance, a feature that can be especially important for enhancing survival of predatory and parasitic species introduced into colder regions for biological control. Such naturally occurring variation, of course, also provides the capacity for pest species to expand their ranges into colder regions.

Crosses between selected lines of *D. melanogaster* suggest that the elements controlling cold hardiness in this species are dispersed over all chromosomes, but chromosome 2 makes the major contribution to cold hardiness in eggs and pupae, while chromosome 3 contributes most to cold hardiness in larval and adult stages (Tucic 1979). Crosses between *D. takahashii* and *D. lutescens* suggest that the genes regulating cold hardiness are located on autosomes (Kimura 1982). One of the acteylcholinesterase mutants of *D. melanogaster*, Ace^{IJ29}, is a conditional mutation that is lethal if the fly is reared at temperatures below 20°C (Greenspan et al. 1980). A simple point mutation that replaces a single serine with a proline is responsible for this effect (Mutero et al. 1994). The mutation alters the secretion rate of acetylcholinesterase, most likely by affecting its folding. This problem is exacerbated by low temperature and results in secretion of an insufficient amount of acetylcholinesterase.

Within the life of a single individual the capacity for cold tolerance also differs from one developmental stage to another. Among nondiapausing individuals of *Sarcophaga crassipalpis,* the stage most tolerant of a cold shock at $-10°C$ is the pupa, followed by pharate adult > adult > larva (Chen et al. 1991b). Interestingly, the stages most tolerant of high temperature stress are also the pupa and pharate adult. The most dramatic developmental differences, however, are associated with diapause. Diapausing pupae of *S. crassipalpis* survive for months at temperatures as low as $-20°C$ (just above their supercooling point), while nondiapausing pupae and other developmental stages are killed by brief exposures to temperatures of $-10°C$ or higher (Adedokun & Denlinger 1984, Lee & Denlinger 1985). Different melanic forms of the same developmental stage of the same species may also have different properties. Body color contributes to body temperature and the rate of warming. Radiant heat is more quickly absorbed by a dark body than by a light body, as illustrated in Fig. 3.3 by the more rapid rate of warming in a melanic form of the ladybird beetle *Adalia bipunctata* than in the non-melanic form of the same species (De Jong et al. 1996). Appreciating the profound differences in cold tolerance associated with different developmental stages and forms is, of course, critical for the design of pest management strategies that exploit low temperature.

Causes of Low Temperature Injury

Many insects do not survive chilling and die due to various forms of non-freezing injury. Although the actual mechanisms responsible for this form of injury remain largely unknown, information from cryobiological investigations using primarily microbial and mammalian cell models provides useful clues. Direct effects of chilling include decreases in the rate of enzymatic activity as well as changes in tertiary structure of proteins and disassembly of polypeptide subunits causing protein denaturation that may be irreversible upon warming (Morris & Clarke 1987). Low temperature induced depolymerization of cytoplasmic microtubules is frequently reported, however this phenomenon has received little attention with respect to insect cold-hardiness.

Nonfreezing injury due to low temperature exposure is frequently associated with damage to the plasma membrane (Steponkus 1984, Drobnis

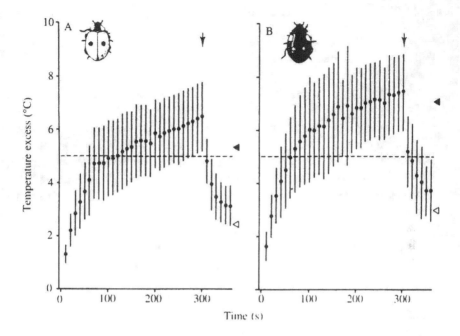

FIGURE 3.3 Warm-up curves at 3 °C for (A) non-melanic and (B) melanic ladybird beetles *Adalia bipunctata*. Mean ± S. D. Values are increases in body temperature after exposing beetles to lights (675Wm⁻²). The vertical arrow indicates the point at which the fan was switched on. The predicted temperature excesses are indicated by the arrowheads on the right (solid, without wind; open, with wind). From DeJong et al. (1996).

et al. 1993, Hazel 1995). At some point chilling induces fluid to gel phase transitions in cell membranes that result in major alterations in membrane permeability, reduction in the activity of membrane bound enzymes, and separation of membrane proteins and lipids into distinct domains that remain even after warming (Quinn 1985, Hazel 1995). Again, few investigations have explored the significance of these effects in insects despite the fact that these membrane related effects have received considerable attention in microorganisms, plants, and lower vertebrates.

To appreciate the nature of freezing injury it is first necessary to consider the dynamics of ice nucleation and freezing within the insect. It is commonly held that survival of freezing at temperatures naturally experienced requires that ice formation be restricted to the extracellular spaces (but see reports describing survival of intracellular freezing in fat body cells by Salt 1959, 1962, Lee et al. 1993b). Initially ice nucleation occurs outside the cells, sometimes seeded by ice nucleating proteins or other nucleators. Because only water molecules can join the growing ice lattice, dissolved solute in the remaining unfrozen body fluids becomes concentrated. This freeze concentration of extracellular fluids causes the osmotic removal of cellular water. As more ice forms, more water leaves the cells.

Although mechanical injury due to internal ice formation can be a deleterious consequence of freezing, excessive concentration of body fluids and cellular dehydration are believed to be the primary stresses (Mazur 1984, Karow 1991). Freeze-concentration may elevate the levels of specific solutes, particularly electrolytes, to the point where they cause protein denaturation and extreme changes in pH. Excessive increases in the osmotic pressure of body fluids may also cause injury. The critical minimum cell volume hypothesis attributes freezing injury to excessive cellular shrinkage that damages the membrane to the point where it is unable to recover upon thawing (Meryman 1974).

Reports from the plant literature suggest that chilling injury may be attributed to oxidative stress (Jahnkhe et al. 1991, Walker & McKensie 1993, Prasad et al. 1994). Injury to the mitochondrial membrane and the proteins involved in electron transport could result in generation of free radicals and other prooxidants. Rojas and Leopold (1996) present intriguing evidence that a similar scenario may be operating in insects. In the house flies they examined, the most cold resistant stages, the pupa and pharate adult, have the highest activity of superoxide dismutase, the scavenging enzyme that represents the first line of defense against oxygen free radicals. Furthermore, they demonstrated elevation of superoxide dismutase activity in response to chilling. Superoxide dismutase converts oxygen free radicals into hydroxyl radicals and hydrogen peroxide, products that are then rendered less toxic to the cell by the action of glutathione. In house flies, the level of this important tripeptide, glutathione, declines during cold storage, further suggesting that oxidative stress may contribute to chilling injury.

Certain systems are more vulnerable to low temperature injury than others. The neuromuscular system appears to be particularly vulnerable. As temperatures decline, insects gradually lose their ability to fly and at slightly lower temperatures they lose their ability to walk. Chill coma, the point at which the insect loses its ability to walk, coincides with the temperature at which the muscles and nerves lose their electrical excitability (Goller & Esch 1990, Xu & Robertson 1994). This point is reached at 12.8 °C in honey bee drones, at 10.6°C in honey bee workers, and at 7°C in adults of *D. melanogaster* (Hosler & Esch 1998). As temperatures drop toward the onset of chill coma several features of the muscle potential change. As shown in the example of honey bee queens (Fig.3.4), the resting potential of the muscle membrane gradually decreases, amplitude of the muscle potential decreases and duration of the muscle potential increases (Hosler et al. 1998). A final burst of muscle potentials is observed just as the insect enters chill coma. The gradual loss of electrical activity is presumed to result from the loss in function of the ion channels needed to maintain the ionic balance essential for generating the potential difference across the membrane.

While the problems associated with brief periods of chill coma are readily reversible, a more severe cold shock can produce nonreversible injury to the neuromuscular system. Flesh flies cold shocked as pharate adults continue to develop, but if the injury is sufficiently severe, adult flies fail to escape from the puparium (Yocum et al. 1994). Tensiometric measurements of eclosion behavior demonstrate that the first signs of injury are reflected in an alteration of the contraction patterns (Fig. 3.5), rather than the intensity of the muscular contractions. This response is in contrast to the impairment observed at high temperature (Chapter 2, Fig. 2.4). With heat shock the patterns of the

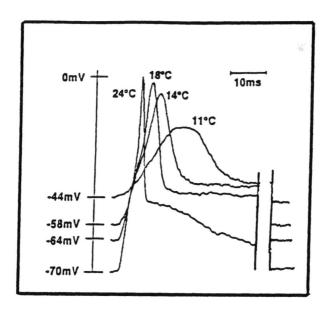

FIGURE 3.4 Temperature effects on the resting potentials and the amplitude and duration of the muscle potentials in thoracic muscles of queen honey bees, *Apis mellifera*. From Hosler et al. (1998).

contractions remained intact long after the intensity of the contractions was diminished. Though both heat shock and cold shock prevent eclosion, the nature of the injury differs. This suggests that the fly is more susceptible to CNS impairment at low temperatures and more susceptible to muscle injury at high temperatures. The circadian gate regulating the precise timing of eclosion within the daily light:dark cycle was altered by heat shock, delayed from dawn to mid-photophase, but no such alterations were observed by cold shock.

The proboscis extension bioassay was also used to evaluate neuromuscular injury in *S. crassipalpis* (Kelty et al. 1996). Adult flies that had been cold shocked as pharate adults fail to extend their proboscis in response to sucrose solutions and fail to groom properly. Cold shock decreases the resting membrane potential in leg muscle fibers and the conductance velocities of the motor neurons innervating the leg muscle. In addition, neuromuscular transmission is impaired as indicated by a lack of evoked end plate potentials (Fig. 3.6). Most likely all of these effects on the neuromuscular system can be traced to disruption in the integrity of the cell membrane, as discussed above.

The reproductive system may be even more vulnerable to low temperature injury. Flesh flies that have been cold shocked as pharate adults may successfully escape from the puparium, feed, mate, but still not reproduce normally. While cold shock injury is less dramatic than heat shock injury on the reproductive processes some impairment is still evident in both males and females: fewer eggs are produced and the fertility rate is lower (our unpublished results). In the house fly, *Musca domestica*, females cold shocked as pharate adults produced fewer eggs during their adult life than

FIGURE 3.5 (Opposite page) Representative tensiometric records of ptilinum movements of eclosing adults of the flesh fly, *Sarcophaga crassipalpis*, that were held at (A) 25°C or were either cold shocked at −10°C for (B) 45 min, (C) 60 min, or (D) 75 min or (E) exposed to 0°C for 10 days as pharate adults. The time scale indicates 10 s intervals; vertical bars indicate a 0.1 mm displacement of the tensiometric sensor. POR, program for obstacle removal, a stereotypic behavior program used for removal of the cap of the puparium and for the removal of obstacles; PFM, program for forward movement, a stereotypic behavior program used to move forward when unobstructed by obstacles. In the least severe cold shock (B), the intensity of the muscular contractions remained strong, but the centrally generated patterns were altered, and the pattern alteration became more pronounced with cold shocks of longer durations (C and D). With long-term chilling (E), the patterns remained intact but the intensity of the muscular contractions decreased. From Yocum et al. (1994).

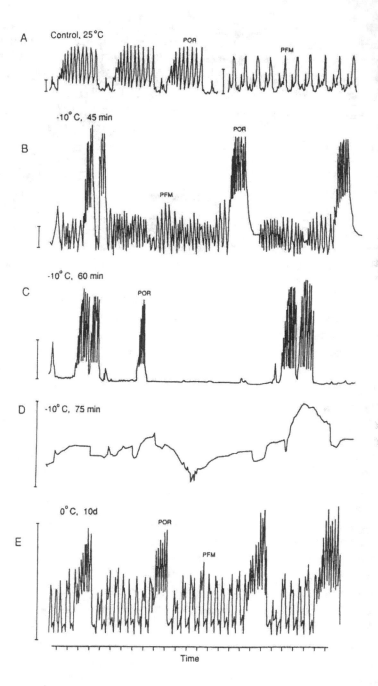

controls that were not cold shocked (Coulson & Bale 1992). Reduced lifetime fecundity was a result both of the female's shorter life span and a reduction in the number of eggs she produced each day. In addition, viability of the eggs produced by the cold shocked females was lower. Similar reductions in fertility caused by cold injury have been reported for other insects including the aphids *Sitobion avenae* (Parish & Bale 1993) and *Rhopalosiphum padi* (Hutchinson & Bale 1994) and the lacewing *Chrysoperla carnea* (Chang et al. 1996).

Cold Hardening

The injury caused by low temperature can frequently be mitigated by prior exposure to less severe low temperatures. Like the acquisition of thermotolerance at high temperature (Chapter 2), cold hardening enables an insect to survive at low temperatures that would otherwise prove lethal. Cold hardening can be either a long term process attained after weeks or months at a low temperature or a very rapid process invoked within minutes or hours after exposure to low temperature.

The traditional view of cold hardening depicts a slow process that gradually increases the insect's low temperature tolerance. This slow acquisition of low temperature tolerance appears to be common in field populations of insects. As temperatures gradually drop in the autumn, overwintering stages become progressively more cold hardy. For example, diapausing pupae of *Sarcophaga bullata* reared outside in central Ohio are not nearly as tolerant of an exposure to -17°C in September or October as they are from November to February (Chen et al. 1991a), and larvae of the goldenrod gall fly, *Eurosta solidaginis*, cannot tolerate -40°C in September or October, but do so in late autumn and winter (Baust & Nishino 1991). For diapausing insects this increase in cold hardiness may be in direct response to low temperature cues, as it is in the European corn borer, *Ostrinia nubilalis* (Hanec & Beck 1960), or it may simply increase with time at a constant temperature, as it does in *S. crassipalpis* (Lee et al. 1987b). Just as cold hardening in a diapausing insect increases gradually over time, it can also gradually decrease over an extended period of time. At the onset of development, a drop in cold hardiness is quite striking. A rapid loss in cold hardiness is usually noted within a few days, but a more subtle decline in cold hardiness is frequently apparent toward the end of diapause, long before the termination of diapause is apparent. Diapausing pupae of *S. crassipalpis*

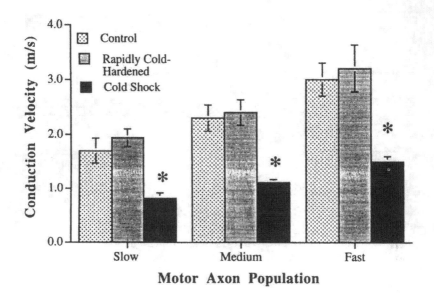

FIGURE 3.6 The effects of cold shock and rapid cold hardening on conduction velocities of the three motor axons (slow, medium, fast) innervating the tergotrochanteral muscle of *Sarcophaga crassipalpis*. For each motor neuron, cold shock was associated with a significant decrease in mean conduction velocity, a decrease which was prevented by rapid cold hardening. From Kelty et al. (1996).

gradually become less tolerant of $-17\,^{\circ}C$ during the 3-4 weeks before they initiate adult development (Lee et al. 1987b). This cold hardening process is thus characterized by both a slow acquisition and a slow decay of tolerance.

Rapid cold hardening, as the name implies, is in marked contrast to the slow, gradual attainment of increased cold tolerance. In this case, the hardening process occurs very fast and enables the insect to quickly respond to low temperature conditions. For example, in *S. crassipalpis*, pharate adults reared at $25\,^{\circ}C$ cannot survive direct exposure to $-10\,^{\circ}C$, but if they are first exposed to $0\,^{\circ}C$ for 10 min or more, they readily survive a 2 h challenge at $-10\,^{\circ}C$ (Fig. 3.7). Rapid cold hardening prevents the neuromuscular damage inflicted by cold shock (Yocum et al. 1994, Kelty et al. 1997). This is the type of response that presumably enables an insect to track daily temperature changes and respond quickly to a drop in temperature. It is not a response restricted to any single developmental stage. Though rapid cold hardening was

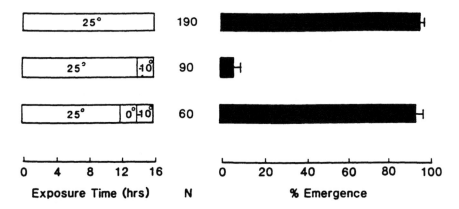

FIGURE 3.7 The effect of cold shock injury (2 h at - 10°C) and the effect of rapid cold hardening (2 h at 0°C) in preventing cold shock injury in the flesh fly, *Sarcophaga crassipalpis*. Flies were tested as pharate adults, and mean ± SE survivorship to the adult stage was evaluated in three replicates of 20 flies each. From Chen et al. (1987a).

first reported in pharate adults of the flesh fly, it has now been reported for a range of species and developmental stages including larvae, diapausing and nondiapausing pupae and pharate adults of *S. bullata* and *S. crassipalpis*, larvae of the thrips *Frankliniella occidentalis* (McDonald et al. 1997), and adults of the elm leaf beetle, *Xanthogaleruca luteola*, the milkweed bug, *Oncopeltus fasciatus* (Lee et al. 1987a), *Drosophila melanogaster* (Czajka & Lee 1990; Chen & Walker 1994), *Musca domestica* (Coulson & Bale 1992), *Culicoides variipennis sonorensis* (Nunamaker 1993), and the monarch butterfly, *Danaus plexippus* (Larsen & Lee 1994). The capacity for rapid cold hardening appears to be a highly conserved trait. In practical terms, the speed of this response underscores its potential for subverting attempts to deliver a lethal cold shock.

The protection that 0°C exposure offers against injury at - 10°C in *S. crassipalpis* reaches a maximum with a 1 day exposure to 0°C but is eventually lost within 20 days if the flies are held continuously at 0°C (Chen & Denlinger 1992). Protection can be extended, however, if the flies held at 0°C receive an intermittent pulse of higher temperature. A one day pulse of 15°C on day 10 extends protection beyond the 20 day period, but pulse temperatures higher than 20°C or lower than 10°C are ineffective. Possibly the 15°C pulse enables the fly to regenerate certain energy resources or cryoprotectants that are progressively depleted at 0°C. A similar success with intermittent pulses was observed with *Musca domestica, Phaenicia sericata,* and *Lucilia cuprina* stored at 10°C and given periodic pulses of 25-28°C (R. A. Leopold & R. R. Rojas, unpublished observation, see Chapter 9). These results suggest the potential for using intermittent pulses of high temperature to sustain low temperature tolerance, a feature that may be especially valuable for maintaining stocks of insects in cold storage. The results also suggest that the natural pattern of temperature cycling may very well play an important role in maintaining low temperature tolerance. But, in contrast, interruption of low temperature exposure (-10 to -15°C) by a 14-day exposure to 2°C did not enhance survival in diapausing pupae of the cabbage root fly, *Delia radicum* (Turnock et al. 1985), nor did interruption of -10°C exposure with periods at 0 or -5°C prevent injury in the bertha armyworm, *Mamestra configurata* (Turnock et al. 1983). Post-stress temperatures, however, play a critical role in survival of diapausing pupae of *M. configurata:* pupae that were cold stressed for 3 days at -14.5°C survived much better at 0°C if they were briefly (1-24h) exposed to 20°C before being held at 0°C (Turnock & Bodnaryk 1993). Survival of cold-stressed pupae at 0°C was much lower for those not given a 20°C pulse. These examples suggest that intermittent pulses as well as post-stress pulses of a higher temperature may be important for both prevention of and recovery from low temperature injury. The precise conditions needed to generate or extend protection are likely to vary considerably with species and developmental stage.

Mechanisms of Cold Hardening

Removal of Ice Nucleators

A key cold hardening mechanism for freezing intolerant species is the removal of efficient internal ice nucleators that would otherwise limit the insect's capacity to supercool. When insects empty their gut during the autumn their capacity to supercool frequently is markedly enhanced (Chapter 4, Cannon & Block 1988). This result suggests that gut contents harbor efficient ice nucleating agents, however in most cases the actual ice nucleating agent has not been identified. A number of freezing intolerant insects have ice nucleating active bacteria as normal flora in the gut (Lee et al. 1991, 1993a). Presumably these bacteria must be removed from the gut or their ice nucleating activity reduced during the winter. Alternatively insects could select protected hibernacula in which environmental temperatures do not decrease below that of their supercooling point.

Water Loss

Not surprisingly, the water relations of insects have a fundamental bearing on cold hardening. Absolute reduction in body water, as has been reported for a variety of overwintering insects (Ring 1982, Zachariassen 1991), decreases the chance of mechanical injury as ice forms in tissues and increases the relative concentration of cryoprotectants by reducing solvent volume. Seasonal changes in the level of "bound" or unfreezable water have been reported (Storey et al. 1981). The nature of this binding remains controversial (Franks 1985) but appears to be associated with interactions between macromolecules and other cellular components (Clegg 1987). Such binding would function to resist cellular water loss to the extracellular space during freezing.

Winter low temperatures are closely tied to the reduced capacity of atmospheric air to carry water vapor and generally lower relative humidities. A recent review emphasized links between cold hardening and resistance to desiccation (Ring & Danks 1994). For example, the accumulation of cryoprotectants in the hemolymph decreases the vapor pressure of supercooled fluids, thereby reducing the gradient promoting water loss to external ice within the hibernaculum (Lundheim & Zachariassen 1993). Although the scientific literature has generally focused on the role of cryoprotectants and water balance for survival at low temperatures, cryoprotective adaptations also confer increased resistance to desiccation stress. A link is also evident between desiccation and cold stress in *Tenebrio molitor* (Kroeker & Walker 1991). In this species a 28 kDA hemolymph protein increases dramatically in response to desiccation stress, and interestingly, also in response to cold stress. How such a protein may function in response to desiccation and cold hardiness remains unknown.

Polyols, Sugars, and Amino Acids

A particularly notable adaptation of overwintering insects is their synthesis and accumulation of exceptionally high concentrations of low-molecular-mass polyols and sugars. Hemolymph levels of these cryoprotectants commonly reach several tenths molar to multimolar levels. Glycerol levels in a larval wasp reached 5M and comprised 25% of its body weight (Salt 1961). Other species, like gall fly larvae of *E. solidaginis*, produce several cryoprotectants (glycerol, sorbitol, trehalose) as they cold harden in preparation for winter (Baust & Lee 1981, Storey & Storey 1981).

These low-molecular-mass polyols and sugars confer increased cold tolerance in several ways. In species that must avoid ice formation in their body fluids the accumulation of cryoprotectants increases their capacity to supercool (Duman et al. 1995). For freezing tolerant species these compounds cause a marked colligative depression (1.86°C per osmole) of the hemolymph melting point. This effect is significant because it reduces the amount of ice that can form at a given sub-zero temperature and consequently decreases cellular dehydration (Karow 1991). Cryoprotectants that penetrate the cell membrane reduce the severity of osmotic gradients generated as ice forms outside the cells and help to retain cytoplasmic water, thereby avoiding excessive cellular dehydration. Cryoprotectants also function to protect cells by stabilizing proteins and cell membranes during freezing and thawing (Carpenter & Crowe 1988, Crowe et al. 1990). However, the accumulation of cryoprotectants does not completely explain the nature of cold hardening at the cellular level. A recent study by Bennett & Lee (1997) using logistic regression modeling revealed that freeze tolerance of *E. solidaginis* fat body cells frozen *in vivo* is consistently greater than for cells frozen *in vitro*, even when cryoprotectants are added to the culture medium.

Hemolymph concentrations of certain free amino acids, most notably alanine and proline, are also frequently elevated in response to low temperature (e.g., Mansingh 1967, Morgan and Chippendale 1983, Fields et al. 1998). Such increases directly correlate with increases in cold tolerance, and it is likely that these free amino acids contribute to cryoprotection. Yet, the precise manner in which this is achieved has not been carefully examined.

Thermal Hysteresis Proteins

Thermal hysteresis refers to a difference between the freezing and melting points of the body fluid. At equilibrium one would expect the freezing and melting points to be nearly identical, but this relationship can be altered by thermal hysteresis proteins (THPs), also known as antifreeze proteins (Duman et al. 1993). THPs depress the freezing point of water by a non-colligative mechanism while leaving the melting point unchanged. In the presence of THPs the freezing point may be lowered 5-6°C below the melting point (Fig. 3.8), thus considerably expanding the organism's low temperature tolerance. THPs were first discovered in cold water, marine fish (DeVries 1971) but were found soon thereafter in a tenebrionid beetle (Duman 1977) and are now known in numerous species of beetles and representatives of many of the lower orders of insects (Duman et al. 1993). THPs appear to be rare in Lepidoptera (Hew et al. 1983) and have not yet been found in Diptera or Hymenoptera.

THPs from several species have been purified and partially characterized (Duman et al. 1993). Molecular masses are in the 14-20 kDa range, and multiple forms of very similar THPs may be present in a single species. Unlike the THPs found in fish, none of the THPs thus far examined in insects contains a carbohydrate component. Maximum activity of the THPs, at least in the beetle *Dendroides canadensis,* is attained when they are bound to a 70 kDa protein in the hemolymph (Wu & Duman 1991).

Synthesis of THPs is a seasonal event. They are produced by the fat body in response to short daylength and low temperature of autumn, persist during the winter and then disappear in response to long daylength in the spring (Fig. 3.8). In larvae of both *D. canadensis* (Horwath & Duman 1983) and *Tenebrio molitor* (Xu et al. 1992) synthesis of THPs is prompted by topical application of juvenile hormone. Changes in the hemolymph titer of juvenile hormone activity are also consistent with the idea that the autumn increase in THPs is mediated, at least partially, by the juvenile hormones.

The utility of THPs for avoiding freezing may have several dimensions. A drop in the freezing point obviously enhances the insect's supercooling capacity, an effect that appears to be achieved by masking ice nucleators

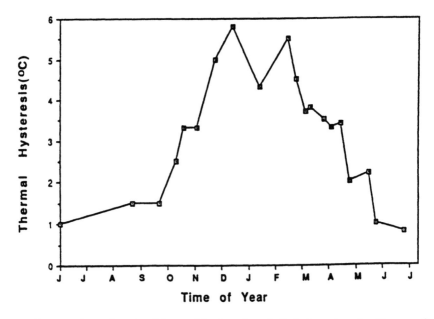

FIGURE 3.8 Annual cycle of thermal-hysteresis activity in hemolymph of larvae of the beetle, *Dendroides canadensis.* From Duman et al. (1993).

present in the hemolymph. In addition, THPs may function to inhibit inoculative freezing by associating with the epidermal cells and thus constructing a barrier to external ice. The presence of THPs in some freeze-tolerant species is a bit more puzzling. It is normally assumed to be advantageous for a freeze-tolerant species to freeze at a relatively high temperature, thus the presence of THPs in such species is unexpected. Yet, THPs are evident in the freeze-tolerant centipede *Lithobius forficatus* where they appear to play a role in protecting the cells from injury during freezing of the extracellular fluids (Tursman & Duman 1995).

Ice Nucleator Proteins

Ice nucleator proteins function in just the opposite manner from thermal hysteresis proteins. Rather than inhibiting freezing, ice nucleator proteins promote freezing. Ice nucleator proteins facilitate the organization of water molecules into embryo crystals which, in turn, "seed" the supercooled solution, causing freezing at relatively high temperatures. As discussed above, elevation of the freezing temperature is advantageous for a freeze-tolerant species, and proteins with this property have now been identified in several insects (Duman et al. 1995).

The best characterized ice nucleator protein is a globular 800 kDa lipoprotein isolated from the hemolymph of the crane fly *Tipula trivittata* (Duman et al. 1985, Neven et al. 1989). This lipoprotein, consisting of 45% protein, 51% lipid, and 4% carbohydrate, contains two apolipoproteins. Unlike most insect lipophorins, this lipoprotein contains phosphatidylinositol, a component deemed essential for ice nucleating activity. An increase in concentration of the lipoprotein yields progressively higher nucleation temperatures, up to a maximum of $-6\,^{\circ}C$ at concentrations at or above $1.7 \times 10^{-7}M$ (Duman et al. 1992). This is possibly due to the fact that individual proteins appear to organize into chains (Yeung et al. 1991), a feature that may increase the availability of nucleation sites.

Stress Proteins

Synthesis of heat shock proteins is a well documented response to high temperature (Chapter 2). Some of the same proteins are synthesized in response to anoxia, heavy metals and other forms of metabolic stress, thus the term stress proteins more accurately captures the diversity of stresses that can stimulate their synthesis. More recently, cold shock was added to the list of stressors capable of stimulating stress protein synthesis (Denlinger et al. 1991). Stress protein synthesis in response to cold shock has been documented in *Drosophila melanogaster* (Burton et al. 1988), *Sarcophaga crassipalpis* (Joplin et al. 1990), the gypsy moth, *Lymantria dispar* (Yocum et al. 1991, Denlinger et al. 1992), and several other insect species.

As with the heat shock response, the most prominent stress protein elicited by cold shock is a member of the heat shock 70 protein family. In *S. crassipalpis* the protein most highly expressed in response to both heat and cold shock is a 72 kDa protein, a protein recognized by an antibody to the 70 kDa heat shock cognate protein in *D. melanogaster* (Joplin et al. 1990). A 92 kDa protein is also synthesized by *S. crassipalpis* in response to both heat shock and cold shock. In addition, several potentially interesting proteins with molecular masses of 78, 45, and 23 kDa are synthesized in the integument, but not the brain, following cold shock. Such cold-shock specific proteins are likely to have special properties unique to the low temperature response. Differences in tissue responses also suggest the complexity inherent in the insect's adaptation to low temperature. The involvement of stress proteins in low temperature responses is not unique to insects. Spinach seedlings acclimated to 5°C also boosted synthesis of proteins in the 70 kDa heat shock family (Neven et al. 1992, Anderson et al. 1994), and like the flies, plants, and bacteria synthesize proteins that are unique to low temperature. One such protein found in *Escherichia coli* (Jones et al. 1987) and *Photorhabdus* sp. (Clarke & Dowds 1994) is a polynucleotide phosphorylase.

Several aspects of stress protein synthesis in response to cold shock differ from the insect's response at high temperature. (1) Synthesis is observed during recovery rather than during the actual stress. A role in the events of recovery is likely. While this implies that stress proteins are unlikely contributors to rapid cold hardening, they may offer protection against subsequent low temperature injury. (2) Synthesis of the stress proteins is concurrent with normal protein synthesis. This is in contrast to the heat shock response. At high temperatures, synthesis of other proteins ceases while stress proteins are being produced. (3) Synthesis can persist for days rather than the brief (minutes or hours) interval of synthesis observed at high temperatures. The duration of the response is especially striking in diapausing pharate larvae of the gypsy moth (Yocum et al. 1991). In this case stress protein synthesis persists for at least 6 days after the cold shock.

The persistence of stress protein expression during gypsy moth diapause (Yocum et al. 1991) suggests a possible role in the cold hardening associated with diapause. In this species the diapausing pharate larvae become cold hardy only after they have been chilled, and it is this period of chilling that capacitates the gypsy moth to synthesize the stress proteins (Denlinger et al. 1992). Our unpublished results with flesh flies also indicate a persistent expression of certain stress proteins during pupal diapause. Both the gypsy moth and flesh flies are freeze intolerant species. The response differs in the goldenrod gall fly, *Eurosta solidaginis,* a freeze tolerant species. Though the gall fly readily synthesizes stress proteins in response to high temperature it does not do so when subjected to low temperature (Lee et al. 1995). Whether this represents a general trend distinguishing freeze-tolerant and freeze-intolerant species awaits validation from additional species.

The function of the stress proteins at low temperature remains unknown, but clearly several functions attributed to members of the 70 kDa heat shock protein family (Craig et al. 1993, Parsell & Lindquist 1993, Morimoto et al. 1994) could prove equally useful at low temperatures. Important enzymes are subject to denaturation at low temperature. Stress proteins could target denatured enzymes for elimination or serve to renature the enzymes. Roles in protein folding, assembly of oligomeric complexes, and chaperoning functions are all known functions for stress proteins and could very well contribute to maintaining cell function at low temperature.

Vitrification

Vitrification of body water is another possible mechanism of cold tolerance that may operate in insects. Vitrification refers to a physical state in which water becomes an amorphous solid or glass. Theoretically vitrification of body water avoids ice nucleation and growth of the ice lattice leading to mechanical injury. In woody plants high concentrations of sugars, particularly sucrose, raffinose, and stachyose, induce vitrification at temperatures as high as -20°C (Chen et al. 1995, Hirsh et al. 1985). Wasylyk et al. (1988) reported partial glass formation in a simulated hemolymph preparation and in intact larvae of *E. solidaginis,* and suggested that this vitrification may provide cryoprotection under natural conditions.

It is thus evident that cold hardening entails a complex suite of responses and can no longer be regarded as a process driven by a single biochemical event such as polyol synthesis. In addition to the mechanisms discussed above, such features as superoxide dismutase activity, glutathione concentrations, energy reserves, and other biochemical parameters are likely to be important contributors to cold hardening. Species differences are likely to dictate that one particular process may be more important in one species than in another, but insects clearly have an array of responses at their disposal, and several mechanisms are likely to operate in any one species at the same time.

Blockage of Cold Hardening

A few studies have investigated ways to prevent or block protective mechanisms of cold hardening. One approach seeks to diminish the natural capacity of freezing intolerant insects to supercool by applying ice nucleating active bacteria and fungi (Fields 1993, Lee et al. 1993a, Chapter 4). These microorganisms are highly efficient ice nucleators that can markedly elevate the supercooling points of a variety of insects. Attempts are currently underway to develop methods using these ice nucleating microorganisms for the control of insect pests.

The capacity to rapidly cold harden is inhibited in *S. crassipalpis* by exposure to anoxic conditions (Yocum & Denlinger 1994). In this study pharate adults that were exposed to 0°C for 2 h prior to a 2-hour period at -10°C survived better than ones directly placed at -10°C. However, this rapid cold hardening at 0°C did not occur under anoxic conditions. This implies the rapid cold hardening that occurs at 0°C is an energy dependent process that can be blocked in the absence of oxygen. Suppressed oxidative metabolism can also prompt the synthesis of anaerobic by-products such as polyols (Wilhelm et al. 1961; Meyer 1980) and other compounds that may function as cryoprotectants. Exposure of the house fly to anoxia while it is within its normal temperature range will indeed stimulate cold hardening (Coulson & Bale 1991), but a similar treatment administered to the flesh fly was ineffective (Kukal et al. 1991). The impact of anoxia thus appears to vary, perhaps with species, developmental status or other experimental conditions. Further research is needed to understand the links between cold shock injury, rapid cold hardening and anoxia. Yet, two groups working with insect pests on cut flowers have effectively coupled exposure to low temperature and hypoxia for quarantine purposes (Seaton & Joyce 1993, Shelton et al. 1996). The advantage, of course, is that the coupling of a low temperature treatment with anoxia may permit the use of a less severe temperature, a feature that is likely to be both less costly as well as less damaging to the fruits or vegetables needing treatment.

Agents that could mask or otherwise incapacitate thermal hysteresis proteins, ice nucleator proteins, stress proteins, interfere with polyol production or other key biochemical processes, or disrupt behavioral responses associated with cold hardening have interesting potential application. The fact that bumblebees can be altered behaviorally by parasitoids to seek cold locations (Müller & Schmid-Hempel 1993) also suggests interesting possibilities for behavioral modification. For diapausing insects numerous tools can be exploited. Invariably the termination of overwintering diapause is associated with a pronounced loss of cold hardiness. By prematurely terminating diapause cold hardiness can also be prematurely lost, thus rendering the insect vulnerable to the low temperatures of winter. Although a diversity of hormonal mechanisms regulate insect diapause (Denlinger 1985), certain patterns are common: many cases of larval diapause can be terminated by a drop in the juvenile hormone titer and/or a pulse of ecdysteroids, pupal diapauses can usually be terminated with ecdysteroids, and most adult diapauses can be broken with juvenile hormone. In addition, diapause in a number of species can be broken with physical manipulations or chemical agents. For example, diapause in flesh flies can be broken by physically shaking the pupae or by exposing pupae to organic solvents such as hexane or ether (Denlinger et al. 1980). While the utility of such tools for breaking diapause have been well demonstrated in the laboratory, few attempts have been made thus far to control pest species with such manipulations.

Future Directions

Insects have a wealth of behavioral and physiological responses to counter the effects of low temperature, and if low temperature is to be used in an effective integrated pest management system, these mechanisms must either be overridden or disabled. The speed of the rapid cold hardening response can quickly subvert attempts to kill the insect if the transfer to low temperature is too gradual. The fact that most insects probably have at their disposal a complex suite of responses suggests a form of double assurance. The insect is not simply relying on a single mechanism for survival but instead involves a complex suite of responses. It is not at all clear how or whether such complex responses are linked. Are the responses somehow integrated through the expression of a master gene, or do distinct cues invoke different aspects of the response?

Overwintering mortality can be extremely high, presumably due both to the low temperatures experienced and to the length of time the insect must depend on energy reserves it has garnered prior to the onset of winter. Low temperatures that prevail during winter are frequently just a few degrees above the insect's lower limit of tolerance. The low temperatures that already prevail during winter thus set the stage for manipulations that subject the insect to a lower temperature (e.g. destruction of its overwintering hibernaculum) or artificially elevate its lower limit of tolerance (e.g. elevation of the supercooling point).

Recent discoveries of ice nucleating bacteria and fungi, thermal hysteresis proteins, ice nucleator proteins, general stress proteins, and cold shock-specific proteins suggest that insects offer a rich source of material for pharmacological prospecting. The enormous diversity of insects suggests that many more such agents or compounds remain to be discovered. Molecular techniques make the small size of insects no longer an obstacle for isolation of interesting, new compounds. Recombinant DNA products that alter freezing or melting points, or offer protection against low temperature injury have potential commercial value as cryoprotective agents in the biomedical field and in agriculture as agents to increase cold tolerance in crops and for increasing the possibilities of cold storage. Transgenic cotton plants that overexpress the superoxide dismutase gene show increased cold tolerance (Allen 1995), and similar manipulations with superoxide dismutase genes or other genes associated with insect cold tolerance may have considerable utility for insects used in biological control or for long term storage of other species.

References

Adedokun, T. A. & D. L. Denlinger. 1984. Cold-hardiness: A component of the diapause syndrome in pupae of the flesh flies, *Sarcophaga crassipalpis* and *S. bullata*. Physiol. Entomol. 9: 361-364.

Allen, R. D. 1995. Dissection of oxidative stress tolerance using transgenic plants. Plant Pysiol. 107: 1049-1054.

Anderson, J. V., Q-B. Li, D. W. Haskell & C. L. Guy. 1994. Structural organization of the spinach endoplasmic reticulum-luninal 70-kilodalton heat-shock cognate gene and expression of 70-kilodalton heat-shock genes during cold acclimation. Plant Physiol. 104: 1359-1370.

Angell, C. A. 1982. Supercooled water, *In* F. Franks, ed. *Water - A Comprehensive Treatise*, pp 1-82. Plenum, New York.

Baker, J. E. 1983. Temperature regulation of larval size and development in *Attagenus megatoma* (Coleoptera: Dermestidae). Ann. Entomol. Soc. Am. 76: 752-756.

Bale, J. S. 1987. Insect cold hardiness: Freezing and supercooling - an ecophysiological perspective. J. Insect Physiol. 33: 899-908.

_____. 1993. Classes of insect cold hardiness. Functional Ecology. 7: 751-753.

Baust, J. G. & R. E. Lee. 1981. Divergent mechanisms of frost-hardiness in two populations of the gall fly, *Eurosta solidaginis*. J. Insect Physiol. 27: 485-490.

Baust, J. G. & M. Nishino. 1991. Freezing tolerance in the goldenrod gall fly, *In* R. E. Lee & D. L. Denlinger, eds. *Insects at Low Temperature*, pp 260-275. Chapman and Hall, New York.

Baust, J. G. & R. R. Rojas. 1985. Review - insect cold hardiness: Facts and fancy. J. Insect Phsiol. 31: 755-759.

Bennett, V. A. & R. E. Lee. 1997. Modeling seasonal changes in intracellular freeze-tolerance of fat body cells of the gall fly *Eurosta solidaginis* (Diptera, Tephritidae). J. Experimental Biol. 200: 185-192.

Block, W. 1991. To freeze or not to freeze? Invertebrate survival of sub-zero temperatures. Functional Ecology 5: 284-290.

Burton, V., H. K. Mitchell, P. Young & N. S. Petersen. 1988. Heat shock protection against cold stress of *Drosophila melanogaster*. Molec. Cell. Biol. 8: 3550-3552.

Cannon, R. J. C. & W. Block. 1988. Cold tolerance of microarthropods. Biological Reviews 63: 23-77.

Carpenter, J. F. & J. H. Crowe. 1988. The mechanism of cryoprotection of proteins by solutes. Cryobiology 25: 244-255.

Chen, T. H. H., M. J. Burke & L. V. Gusta. 1995. Freezing tolerance in plants: An overview, *In* R. E. Lee, G. J. Warren & L. Gusta, eds. *Biological Ice Nucleation and Its Applications*. pp 115-135. American Phytopathology Society, St. Paul, Minnesota.

Chen, C.-P. & D. L. Denlinger. 1992. Reduction of cold injury in flies using an intermittent pulse of high temperature. Cryobiology 29: 138-143.

Chen, C.-P., D. L. Denlinger & R. E. Lee. 1987a. Cold-shock injury and rapid cold hardening in the flesh fly *Sarcophaga crassipalpis*. Physiol. Zool. 60: 297-304.

_____. 1987b. Responses of nondiapausing flesh flies (Diptera: Sarcophagidae) to low rearing temperatures: Developmental rate, cold tolerance, and glycerol concentrations. Ann. Entomol. Soc. Am. 80: 790-796.

_____. 1991a. Seasonal variation in generation time, diapause and cold hardiness in a central Ohio population of the flesh fly, *Sarcophaga bullata*. Ecol. Entomol. 16: 155-162.

Chen, C.-P., R. E. Lee & D. L. Denlinger. 1990. A comparison of the response of tropical and temperate flies (Diptera: Sarcophagidae) to cold and heat stress. J. Comparative Physiol. B 160: 543-547.

_____. 1991b. Cold shock and heat shock: A comparison of the protection generated by brief pretreatment at less severe temperatures. Physiol. Entomol. 16: 19-26.

Chen, C.-P. & V. K. Walker. 1994. Cold shock and chilling tolerance in *Drosophila*. J. Insect Physiol. 40: 661-669.

Chang, Y. F., M. J. Tauber & C. A. Tauber. 1996. Reproduction and quality of F_1 offspring in *Chrysoperla carnea:* Differential influence of quiescence, artificially-induced diapause, and natural diapause. J. Insect Physiol. 42: 521-528.

Clarke, D. J. & B. C. A. Dowds. 1994. The gene coding for polynucleotide phosphorylase in *Photorhabdus* sp. strain K122 is induced at low temperatures. J. Bacteriol. 176: 3775-3784.

Clegg, J. S. 1987. Cytoplasmic organization and the properties of cell water: Speculations on animal cell cryopreservation, *In* D. E. Pegg & A. M. Karow, eds. *The Biophysics of Organ Cryopreservation.* pp 79-81. Plenum, New York.

Costanzo, J. P. & R. E. Lee. 1995. Supercooling and ice nucleation in vertebrate ectotherms, *In* R. E. Lee, G. J. Warren & L. V. Gusta, eds. *Biological Ice Nucleation and Its Applications,* pp 221-237. The American Phytopathological Society, St. Paul, Minnesota.

Costanzo, J. P. & R. E. Lee. 1996. Mini-review: Ice nucleation in freeze-tolerant vertebrates. Cryo-Letters. 17: 111-118.

Coulson, S. J. & J. S. Bale. 1991. Anoxia induces rapid cold hardening in the housefly *Musca domestica* (Diptera: Muscidae). J. Insect Physiol. 37: 497-501.

_____. 1992. Effect of rapid cold hardening on reproduction and survival of offspring in the housefly *Musca domestica.* J. Insect Physiol. 38: 421-424.

Craig, E. A., B. D. Gambill & R. J. Nelson. 1993. Heat shock proteins: Molecular chaperones of protein biogenesis. Microbiol. Rev. 57: 402-414.

Crowe, J. H., J. F. Carpenter, L. M. Crowe & T. J. Anchordoguy. 1990. Are freezing and dehydration similar stress vectors? A comparison of modes of interaction of stabilizing solutes with biomolecules. Cryobiology 27: 219-231.

Cymborowski, B. 1988. Effect of cooling stress on endocrine events in *Galleria mellonella, In* F. Sehnal, A. Zabza and D. L. Denlinger, eds. *Endocrinological Frontiers in Physiological Insect Ecology.* pp 203-212. Technical University of Wroclaw, Wroclaw.

Cymborowski, B. & M. I. Bogus. 1976. Juvenilizing effect of cooling stress on *Galleria mellonella.* J. Insect Physiol. 22: 669-672.

Czajka, M. C. & R. E. Lee. 1990. A rapid cold-hardening response protecting against cold shock injury in Drosophila melanogaster. J. Experimental Biol. 148: 245-254.

Danks, H. V. 1981. *Arctic Arthropods.* Entomol. Soc. Canada, Ottawa.

_____. 1987. *Insect Dormancy: An Ecological Perspective.* Biological Survey of Canada, Ottawa.

_____. 1991. Winter habitats and ecological adaptations for winter survival, *In* R. E. Lee & D. L. Denlinger, eds. *Insects at Low Temperature.* pp 231-259. Chapman and Hall, New York.

David, J. R., R. Allemand, J. Van Herrewege & Y. Cohet. 1983. Ecophysiology: abiotic factors, *In* M. Ashburner, H. L. Carson & J. N. Thompson, eds. *The Genetics and Biology of Drosophia.* Vol. 3d, pp 106-169. Academic Press, London.

De Jong, P. W., S. W. W. Gussekloo & P. M. Brakefield. 1996. Differences in thermal balance, body temperature and activity between non-melanic two-spot ladybird beetles (*Adalia bipunctata*) under controlled conditions. J. Experimental Biol. 199: 2655-2666.

Denlinger, D. L. 1985. Hormonal control of insect diapause, *In* G. A Kerkut & L. I. Gilbert, eds. *Comprehensive Insect Physiology, Biochemistry and Pharmacology.* Vol. 8, pp 353-412. Pergamon, Oxford.

_____. 1986. Dormancy in tropical insects. Annual Review Entomol. 31: 239-264.

_____. 1991. Relationship between cold hardiness and diapause, *In* R. E. Lee & D. L. Denlinger, eds. *Insect at Low Temperature.* pp 174-198. Chapman and Hall, New York.

Denlinger, D. L., J. J. Campbell & J. Y. Bradfield. 1980. Stimulatory effect of organic solvents on initiating development in diapausing pupae of the flesh fly *Sarcophaga crassipalpis* and the tobacco hornworm *Manduca sexta.* Physiol. Entomol. 5: 7-15.

Denlinger, D. L., K. H. Joplin, C.-P. Chen & R. E. Lee. 1991. Cold shock and heat shock, *In* R. E. Lee & D. L. Denlinger, eds. *Insects at Low Temperature.* pp 131-148. Chapman and Hall, New York.

Denlinger, D. L., R. E. Lee, G. D. Yocum & O. Kukal. 1992. Role of chilling in the acquisition of cold tolerance and the capacitation to express stress proteins in diapausing pharate larvae of the gypsy moth, *Lymantria dispar.* Arch. Insect Biochem. Physiol. 21: 271-280.

Denlinger, D. L. & J. Žďárek. 1991. Commitment to metamorphosis in tsetse (*Glossina morsitans centralis*): temporal, nutritional and hormonal aspects of the decision. J. Insect Physiol. 37: 333-338.

DeVries, A. L. 1971. Glycoproteins as biological antifreeze agents in antarctic fishes. Science 172: 1152-1155.

Drobnis, E. Z., L. M. Crowe, T. Berger, T. J. Anchordoguy, J. W. Over street & J. H. Crowe. 1993. Cold shock damage is due to lipid phase transitions in cell membranes: A demonstration using sperm as a model. J. Experimental Biol. 265: 432-437.

Duman, J. G. 1977. The role of macromolecular antifreeze in the darkling beetle *Meracantha contracta.* J. Comparative Physiol. B 115: 279-286.

Duman, J. G., L. G. Neven, J. M. Beals, K. O. Olson & F. J. Castellino. 1985. Freeze tolerance adaptations, including haemolymph protein and lipoprotein ice nucleators, in larvae of the cranefly *Tipula trivittata.* J. Insect Physiol. 31: 1-9.

Duman, J. G., T. M. Olsen, K. L. Yeung & F. Jerva. 1995. The roles of ice nucleators in cold tolerant invertebrates, *In* R. E. Lee, G. J. Warren & L. V. Gusta, eds. *Biological Ice Nucleation and Its Applications.* pp 201-219. American Phytopathological Society, St. Paul, Minnesota.

Duman, J. G., D. W. Wu, T. M. Olsen, M. Urratia & D. Tursman. 1993. Thermal-hysteresis proteins. Advances Low Temperature Biology 2: 131-182.

Duman, J. G., D. W. Wu & K. L. Yeung. 1992. Hemolymph proteins involved in the cold tolerance of terrestrial arthropods: Antifreeze and ice nucleator proteins, *In*

G. N. Somero, C. B. Osmond & C. L. Bolis, eds. *Water and Life: Comparative analysis of Water Relationships at the Organismic, Cellular and Molecular Level.* pp 282-300, Springer-Verlag, Berlin.

Fields, P. G. 1993. Reduction of cold tolerance of stored-product insects by ice-nucleating-active bacteria. Environ. Entomol. 22: 470-476.

Fields, P. G., F. Fleurat-Lessard, L. Lavenseau, G. Febvay, L. Peypelut & G. Bonnot. 1998. The effect of cold acclimation and deacclimation on cold tolerance, trehalose and free amino acid levels in *Sitophilus granarius* and *Cryptolestes ferrugineus* (Coleoptera). J. Insect Physiol. (in press).

Franks, F. 1985. *Biophysics and Biochemistry at Low Temperatures.* Cambridge University Press, Cambridge.

_____. 1987. Ice nucleation and freezing in undercooled cells. Cryobiology 20: 298-309.

Gierke, E. von. 1932. Über die Hautungen und die Entwicklungsgeschwindigkeit der Larven der Mehlmotte *Ephestia kuhniella* Zell. Roux' Arch. 127: 387-410.

Goller, F. & H. Esch. 1990. Comparative study of chill-coma temperatures and muscle potentials in insect flight muscles. J. Experimental Biol. 150: 221-231.

Gösswald, K. & K. Bier. 1955. Beeinflussung der Geschlechtsverhaltnisse durch Temperatureinwirkung bei *Formica rufa* L. Naturwissenschaften 42: 133-134.

Greenspan, R. J., J. A. Finn, Jr. & J. C. Hall. 1980. Acetylcholinesterase mutants in *Drosophila* and their effects on the structure and function of the central nervous system. J. Comparative Neurol. 189: 741-774.

Hanec, W. & S. D. Beck. 1960. Cold hardiness in the European corn borer, *Pyrausta nubilalis* (Hubn.). J. Insect Physiol. 5: 169-180.

Hardie, J. & A. D. Lees. 1985. Endocrine control of polymorphism and polyphenism, In G. A. Kerkut & L. I. Gilbert, eds., *Comprehensive Insect Physiology, Biochemistry and Pharmacology.* Vol. 8, pp. 441-490. Pergamon, Oxford.

Hazel, J. R. 1995. Thermal adaption in biological membranes: Is homeoviscous adaptation the explanation? Annual Review Physiol. 57: 19-42.

Heinrich, B. 1987. Thermoregulation by winter-flying endothermic moths. J. Experimental Biol. 127: 313-332.

Hew, C. L., M. H. Kao & Y. P. So. 1983. Presence of cystine-containing antifreeze proteins in the spruce budworm, *Choristoneura fumiferana.* Can. J. Zool. 61: 2324-2328.

Hirsh, A. G., R. J. Williams & H. T. Meryman. 1985. A novel method of natural cryopreservation: Intracellular glass formation in deeply frozen *Populus.* Plant Physiology. 79: 41-56.

Hoelscher, C. E. & S. B. Vinson. 1971. The sex ratio of a hymenopterous parasitoid, *Campoletis perdistinctus*, as affected by photoperiod, mating, and temperature. Ann. Entomol. Soc. Amer. 64: 1373-1376.

Hoffmann, A. A. & M. Watson. 1993. Geographical variation in the acclimation responses of *Drosophila* to temperature extremes. Amer. Naturalist 142: S93-S113.

Horwath, K. L. & J. G. Duman. 1983. Induction of antifreeze protein production by juvenile hormone in larvae of the beetle *Dendroides canadensis*. J. Comparative Physiol. B 151: 233-240.

Hosler, J. S., J. E. Burns & H. E. Esch. 1998. Evaluation of chill-coma, resting potentials and muscle potentials in the queen honeybee *Apis mellifera*. (In preparation).

Hosler, J. S. & H. Esch. 1998. The effect of resting potential on species-specific differences in chill-coma. (In preparation).

Hutchinson, L. A. & J. S. Bale. 1994. Effects of sublethal cold stress on the aphid *Rhopalosiphum padi*. J. Applied Ecol. 31: 102-108.

Jahnkhe, L. S., M. R. Hull & S. P. Long. 1991. Chilling stress and oxygen metabolizing enzymes in *Zea mays* and *Zea diploperennis*. Plant Cell Environ. 14: 97-104.

Jones, P. G., R. A. Van Bogelen & F. C. Neidhart. 1987. Induction of proteins in response to low temperature in *Escherichia coli*. J. Bacteriol. 169: 2092-2095.

Joplin, K. H., G. D. Yocum & D. L. Denlinger. 1990. Cold shock elicits expression of heat shock proteins in the flesh fly, *Sarcophaga crassipalpis*. J. Insect Physiol. 36: 825-834.

Karow, A. M. 1991. Chemical cryoprotection of metazoan cells. BioScience 41: 155-160.

Kelty, J. D., K. A. Killian & R. E. Lee. 1996. Cold shock and rapid cold-hardening of pharate adult flesh flies (*Sarcophaga crassipalpis*): Effects on behavior and neuromuscular function following eclosion. Physiol. Entomol. 21: 283-288.

Kimura, M. T. 1982. Inheritance of cold hardiness and sugar contents in two closely related species, *Drosophila takahashii* and *D. lutescens*. Jap. J. Genetics 57: 575-580.

Kohshima, S. 1984. A novel cold-tolerant insect found in a Himalayan glacier. Nature 310: 225-227.

Kroeker, E. M. & V. K. Walker. 1991. Dsp28: A desiccation stress protein in *Tenebrio molitor* hemolymph. Arch. Insect Biochem. Physiol. 17: 169-182.

Kukal, O., D. L. Denlinger & R. E. Lee. 1991. Developmental and metabolic changes induced by anoxia in diapausing and non-diapausing flesh fly pupae. J. Comparative Physiol. B 160: 683-689.

Larsen, K. J. & R. E. Lee. 1994. Cold tolerance including rapid cold-hardening and inoculative freezing in migrant monarch butterflies in Ohio. J. Insect Physiol. 40: 859-864.

Lauge, G. 1985. Sex determination: genetic and epigenetic factors, *In* G. A. Kerkut & L. I. Gilbert, eds. *Comprehensive Insect Physiology, Biochemistry and Pharmacology*. Vol. 1, pp. 295-318. Pergamon, Oxford.

Leather, S. R., K. F. A. Walters & J. S. Bale. 1993. *The Ecology of Insect Overwintering*. Cambridge University Press, Cambridge.

Lee, R. E. 1989. Insect cold-hardiness: To freeze or not to freeze? BioScience 39: 308-313.

_____. 1991. Principles of insect low temperature tolerance, *In* R. E. Lee & D. L. Denlinger, eds. *Insects at Low Temperature,* pp 17-46. Chapman and Hall, New York.

Lee, R. E., C.-P. Chen & D. L. Denlinger. 1987a. A rapid cold-hardening process in insects. Science 238: 1415-1417.

Lee, R. E., C.-P. Chen, M. H. Meacham & D. L. Denlinger. 1987b. Ontogenetic patterns of cold-hardiness and glycerol production in *Sarcophaga crassipalpis.* J. Insect Physiol. 33: 587-592.

Lee, R. E., J. P. Costanzo & J. A. Mugnano. 1996. Regulation of supercooling and ice nucleation in insects. European J. Entomology 93: 405-418.

Lee, R. E. & D. L. Denlinger, eds. 1991. *Insects at Low Temperature.* Chapman and Hall, New York. 513 pp.

Lee, R. E. & D. L. Denlinger. 1985. Cold tolerance in diapausing and non-diapausing stages of the flesh fly, *Sarcophaga crassipalpis.* Physiol. Entomol. 10: 309-315.

Lee, R. E., A. Dommel, K. H. Joplin & D. L. Denlinger. 1995. Cryobiology of the freeze-tolerant gall fly *Eurosta solidaginis:* Overwintering energetics and heat shock proteins. Climate Res. 5: 61-67.

Lee, R. E., M. R. Lee & J. M. Strong-Gunderson. 1993a. Insect cold-hardiness and ice nucleating active microorganisms including their potential use for biological control. J. Insect Physiol. 39: 1-12.

Lee, R. E. & E. A. Lewis. 1985. Effect of temperature and duration of exposure on tissue ice formation in the gall fly, *Eurosta solidaginis* (Diptera, Tephritidae). Cryo-Letters. 6: 24-34.

Lee, R. E., J. J. McGrath, R. T. Morason & R. M. Taddeo. 1993b. Survival of intracellular freezing, lipid coalescence and osmotic fragility in fat body cells of the freeze-tolerant gall fly *Eurosta solidaginis.* J. Insect Physiol. 39: 445-450.

Lee, R. E., M. M. Strong-Gunderson, M. R. Lee, K. S. Grove & T. J. Riga. 1991. Isolation of ice nucleating active bacteria from insects. J. Experimental Zool. 257: 124-127.

Lengerken, H. von. 1932. Nachinkende Entwicklung und ihre Folgeerscheinungen beim Mehlkafer. Jena Z. Naturw. 67: 260-274.

Lundheim, R. & K. E. Zachariassen. 1993. Water balance of over-wintering beetles in relation to strategies for cold tolerance. J. Comparative Physiol. B. 163: 1-4.

Mansingh, A. 1967. Changes in the free amino acids of the haemolymph of Antheraea pernyi during induction and termination of diapause. J. Insect Physiol. 13: 1645-1655.

Masaki, S. 1980. Summer diapause. Annual Review Entomol. 25: 1-25.

Mazur, P. 1984. Freezing of living cells: Mechanisms and implications. American J. Physiol. 247: c125-c142.

McDonald, J. R., J. S. Bale & K. F. A. Walters. 1997. Rapid cold hardening in the western flower thrips *Frankliniella occidentalis.* J. Insect Physiol. 43: 759-766.

Meryman, H. T. 1974. Freezing injury and its prevention in living cells. Annual Review Biophysics Bioengineering. 3: 341-363.

Meyer, S. G. E. 1980. Studies on anaerobic glucose and glutamate metabolism in larvae of *Callitroga macellaria*. Insect Biochem. 10: 449-455.

Miller, L. K. 1978. Freezing tolerance in relation to cooling rate in an adult insect. Cryobiology 15: 345-349.

Miya, K. & Y. Kobayashi. 1974. The embryonic development of *Atrachya menetriesi* Faldermann (Coleoptera, Chrysomelidae). II. Analysis of early development ligation and low temperature treatment. J. Fac. Agri. Iwate Univ. 12: 39-55.

Morgan, T. D. & G. M. Chippendale. 1983. Free amino acids of the haemolymph of the southwestern corn borer and the European corn borer in relation to their diapause. J. Insect Physiol. 29: 735-740.

Morimoto, R. I., A. Tissieres & C. Georgopoulos, eds. 1994. *The Biology of Heat Shock Proteins and Molecular Chaperones*. Cold Spring Harbor Laboratory Press, New York.

Morris, G. J. & A. Clarke. 1987. Cells at low temperature, *In* B. W. W. Grout & G. J. Morris, eds. *Effects of Low Temperatures on Biological Systems*, pp 72-119. Edward Arnold, Baltimore.

Morrissey, R. & J. S. Edwards. 1979. Neural function in an alpine grylloblattid: A comparison with the house cricket, *Acheta domesticus*. Physiol. Entomol. 4: 241-250.

Mugnano, J. A., R. E. Lee & R. T. Taylor. 1996. Fat body cells and calcium phosphate spherules induce ice nucleation in the freeze-tolerant larvae of the gall fly *Eurosta solidaginis* (Diptera, Tephritidae). J. Experimental Biol. 199: 465-471.

Müller, C. B. & P. Schmid-Hempel. 1993. Exploitation of cold temperature as defense against parasitoids in bumblebees. Nature 363: 65-67.

Munger, F. 1963. Factors affecting growth and multiplication of the citrus red mite, *Panonychus citri*. Ann. Entomol. Soc. Amer. 56: 867-874.

Mutero, A., J.-M. Bride, M. Pralovorio & D. Fournier. 1994. *Drosophila melanogaster* acetylcholinesterase: Identification and expression of two mutations responsible for cold- and heat-sensitive phenotypes. Mol. Gen. Genet. 243: 699-705.

Neven, L. G., J. G. Duman, M. G. Low, L. C. Sehl & F. J. Castellino. 1989. Purification and characterization of an insect hemolymph lipoprotein ice nucleator: Evidence for the importance of phosphatidylinositol and apolipoprotein in the ice nucleator activity. J. Comparative Physiol. 159: 71-82.

Neven, L. G., D. W. Haskell, C. L. Guy, N. Denslow, P. A. Klein, L. G. Green and A. Silverman. 1992. Association of 70-kilodalton heat-shock cognate proteins with acclimation to cold. Plant Physiol. 99: 1362-1369.

Nijihout, H. F. & C. M. Williams. 1974. Control of molting and metamorphosis in the tobacco hornworm, *Manduca sexta* (L.): Growth of the last-instar and the decision to pupate. J. Experimental Biol. 61: 481-491.

Nunamaker, R. A. 1993. Rapid cold-hardening in *Culicoides variipennis sonorensis* (Diptera: Ceratopogonidae). J. Med. Entomol. 30: 913-917.

Parish, W. E. G. & J. S. Bale. 1993. Effects of brief exposures to low temperature on the development, longevity and fecundity of the grain aphid *Sitobion avenae* (Hemiptera: Aphididae). Annals Applied Biol. 122: 9-21.

Parsell, D. A. & S. Lindquist. 1993. The function of heat-shock proteins in stress tolerance: Degradation and reactivation of damaged proteins. Annual Review Genetics 27: 437-496.

Patterson, J. L. & J. G. Duman. 1978. The role of the thermal hysteresis factor in *Tenebrio molitor* larvae. J. Experimental Biol. 74: 37-45.

Prasad, T. K., M. D. Anderson, B. A. Martin & C. R. Stewart. 1994. Evidence or chilling-induced oxidative stress in maize seedlings and a regulatory role for hydrogen peroxide. The Plant Cell 6: 65-74.

Pullin, A. S. 1996. Physiological relationships between insect diapause and cold tolerance: Coevolution or coincidence? European J. Entomol. 93: 121-129.

Quinn, P. J. 1985. A lipid-phase separation model of low-temperature damage to biological membranes. Cryobiology 22: 128-146.

Ring, R. A. 1982. Freezing-tolerant insects with low supercooling points. Comparative Biochem. Physiol. A. 73: 605-612.

Ring, R. A. & H. V. Danks. 1994. Desiccation and cryoprotection: Overlapping adaptations. Cryo-Letters 15: 181-190.

Robertson, F. W. 1957. Studies in quantitative inheritance. XI. Genetic and environmental correlation between body size and egg production in *Drosophila melanogaster*. J. Genet. 55: 428-443.

Rojas, R. R. & R. A. Leopold. 1996. Chilling injury in the house fly: Evidence for the role of oxidative stress between pupariation and emergence. Cryobiology 33: 447-458.

Salt, R. W. 1959. Survival of frozen fat body cells in an insect. Nature 193: 1426.

_____. 1961. Principles of insect cold-hardiness. Annual Review Entomol. 6: 55-74.

_____. 1962. Intracellular freezing in insects. Nature 193: 1207-1208.

Saunders, D. S. 1982. *Insect Clocks,* 2nd ed., Pergamon, Oxford.

Seaton, K. A. & D. C. Joyce. 1993. Effects of low temperature and elevated CO_2 treatments and of heat treatments for insect disinfestation on some native Australian cut flowers. Scientia Horticulturae 56: 119-133.

Sehnal, F. 1991. Effects of cold on morphogenesis, *In* R. E. Lee and D. L. Denlinger, eds., *Insects at Low Temperature*. pp. 149-173, Chapman and Hall, New York.

Sehnal, F. & H. Rembold. 1985. Brain stimulation of juvenile hormone production in insect larvae. Experientia 41: 684-685.

Sehnal, F. & H. Schneiderman. 1973. Action of the corpora allata and of juvenilizing substances on the larval-pupal transformation of *Galleria mellonella* L. (Lepidoptera). Acta Entomol. Bohemoslov. 70: 289-302.

Seiler, J. 1920. Geschlectschromosomenuntersuchungen an Psychiden I. Experimentelle Beeinflussung der geschlectsbestimmenden Reifenteilungen bei *Talaeporia tubulosa*. Arch. Zellforsch. 15: 249-268.

Shelton, M. D., V. R. Walter, D. Brandl & V. Mendez. 1996. The effects of refrigerated, controlled-atomosphere storage during marine shipment on insect mortality and cut flower vase life. HortTechnology 6: 247-250.

Sømme, L. & W. Block. 1991. Adaptations to alpine and polar environments in insects and other terrestrial arthropods, *In* R. E. Lee & D. L. Denlinger, eds. *Insects at Low Temperature.* pp 318-359. Chapman and Hall, New York.

Stellwaag-Kittler, F. 1954. Zur Physiologie der Kaferhautung. Untersuchungen am Mehlkafer *Tenebrio molitor* L. Biol. Zbl. 73: 12-49.

Steponkus, P. L. 1984. Role of the plasma membrane in freezing injury and cold acclimation. Annual Review Plant Physiol. 35: 543-584.

Storey, K. B., J. G. Baust & P. Buescher. 1981. Determination of water "bound" by soluble subcellular components during low-temperature acclimation in the gall fly larva, *Eurosta solidaginis.* Cryobiology 18: 315-321.

Storey, K. B. & J. Storey. 1981. Biochemical strategies of overwintering in the gall fly larva, *Eurosta solidaginis:* Effect of low temperature acclimation on the activities of enzymes of intermediary metabolism. J. Comparative Physiol. 144: 191-199.

Tauber, M. J., C. A. Tauber & S. Masaki. 1986. *Seasonal Adaptations in Insects.* Oxford University Press, Oxford.

Tucic, N. 1979. Genetic capacity for adaptation to cold resistance at different developmental stages of *Drosophila melanogaster.* Evolution 33: 350-358.

Turnock, W. J. & R. P. Bodnaryk. 1991. Latent cold injury and its conditional expression in the bertha armyworm, *Mamestra configurata* (Noctuidae: Lepidoptera). Cryo-Letters 12: 377-384.

_____. 1993. The reversal of cold injury and its effect on the response to subsequent cold exposures. Cryo-Letters 14: 251-256.

Turnock, W. J., T. H. Jones & P. M. Reader. 1985. Effects of cold stress during diapause on the survival and development of *Delia radicum* (Diptera: Anthomyiidae) in England. Oecologia 67: 506-510.

Turnock, W. J., R. J. Lamb & R. P. Bodnaryk. 1983. Effects of cold stress during pupal diapause on the survival and development of *Mamestra configurata* (Lepidoptera: Noctuidae). Oecologia 56: 185-192.

Tursman, D. & J. G. Duman. 1995. Cryoprotective effects of thermal hysteresis protein on survivorship of frozen gut cells from the freeze-tolerant centipede *Lithobius forficatus.* J. Experimental Zool. 272: 249-257.

Vali, G. 1995. Principles of ice nucleation, *In* R. E. Lee, G. J. Warren & L. V. Gusta, eds. *Biological Ice Nucleation and Its Applications.* pp 1-28. American Phytopathological Society, St. Paul, Minnesota.

Villee, C. A. 1943. Phenogenetic studies of the homoeotic mutants of *Drosophila melanogaster* I. The effects of temperature on the expression of aristapedia. J. Experimental Zool. 93: 75-98.

_____. 1944. Phenogenetic studies of the homoeotic mutants of *Drosophila melanogaster* II. The effects of temperature on the expression of proboscipedia. J. Experimental Zool. 96: 85-102.

Walker, M. A. & B. D. McKensie. 1993. Role of ascorbate-glutathione antioxidant system in chilling stress. HortScience 17: 173-186.

Wasylyk, J. M., A. Tice & J. G. Baust. 1988. Partial glass formation: A novel mechanism of insect cryoprotection. Cryobiology 25: 251-258.

Wilhelm, R. C., H. A. Schneidermann & L. J. Daniel. 1961. The effects of anaerobiosis on the giant silkworms *Hyalophora cecropia* and *Samia cynthia* with special reference to the accumulation of glycerol and lactic acid. J. Insect Physiol. 7: 273-288.

Wilson, F. & L. T. Woolcock. 1960. Environmental determination of sex in a parthenogenetic parasite. Nature 186: 99-100.

Wrensch, D. L. 1993. Evolutionary flexibility through haploid males or how chance favors the prepared genome, *In* D. L. Wrensch & M. A. Ebbert, eds. *Evolution and Diversity of Sex Ratio in Insects and Mites,* pp 118-149, Chapman and Hall, New York.

Wu, D. W. & J. G. Duman. 1991. Activation of antifreeze proteins from the beetle *Dendroides canadensis.* J. Comparative Physiol. B 161: 279-283.

Xu, H., J. G. Duman, W. G. Goodman & D. W. Wu. 1992. A role for juvenile hormone in induction of antifreeze protein production by the fat body in the beetle *Tenebrio molitor.* Comparative Biochem. Physiol. 101B: 105-109.

Xu, H. & R. M. Robertson. 1994. Effects of temperature on properties of flight neurons in locust. J. Comparative Physiol. A 175: 193-202.

Yeung, K. L., E. E. Wolf & J. G. Duman. 1991. A scanning tunneling microscopy study of an insect lipoprotein ice nucleator. J. Vac. Sci. Technol. B 9: 1197-1201.

Yocum, G. D. & D. L. Denlinger. 1994. Anoxia blocks thermotolerance and the induction of rapid cold hardening in the flesh fly, *Sarcophaga crassipalpis.* Physiol. Entomol. 19: 152-158.

Yocum, G. D., K. H. Joplin & D. L. Denlinger. 1991. Expression of heat shock proteins in response to high and low temperature extremes in diapausing pharate larvae of the gypsy moth, *Lymantria dispar.* Arch. Insect Biochem. Physiol. 18: 239-249.

Yocum, G. D., J. Ždárek, K. H. Joplin, R. E. Lee, D. C. Smith, K. D. Manter & D. L. Denlinger. 1994. Alteration of the eclosion rhythm and eclosion behavior in the flesh fly, *Sarcophaga crassipalpis,* by low and high temperature stress. J. Insect Physiol. 40: 13-21.

Zachariassen, K. E. 1985. Physiology of cold tolerance in insects. Physiol. Rev. 65: 799-832.

_____. 1991. The water relations of overwintering insects, *In* R. E. Lee & D. L. Denlinger, eds. *Insects at Low Temperature,* pp 47-63. Chapman and Hall, New York.

_____. 1992. Ice nucleating agents in cold-hardy insects, *In* G. N. Somero, C. B. Osmond & C. L. Bolis, eds. *Water and Life,* pp 261-281. Springer-Verlag, Berlin.

Zaslavski, V. A. 1988. *Insect Development, Photoperiodic and Temperature Control.* Springer-Verlag, Berlin.

Žďárek, J. & K. Sláma. 1972. Supernumerary larval instars in cyclorrhaphous Diptera. Biol. Bull. 142: 350-357.

4

Reducing Cold-Hardiness
of Insect Pests Using
Ice Nucleating Active Microbes

Richard E. Lee, Jr., Jon P. Costanzo, and Marcia R. Lee

As concerns continue to mount regarding the environmental and human health consequences of using chemical controls for insect pests, a wide variety of alternative approaches are receiving increased attention. Crop rotation, tillage practices, genetically-engineered crop varieties, and the use of predators, parasites, and pathogens as agents of biological control are representative of these strategies. In this chapter we describe the initial results of research that may lead to a novel strategy for the control of insect pests that naturally overwinter in exposed sites or whose environment can be artificially cooled. This approach relies on the use of ice nucleating active microorganisms to increase the likelihood that pests will experience lethal internal freezing.

Various facets of insect cold-hardiness and overwintering biology have been the subjects of a rather large number of recent reviews and books that range in focus from the biochemical and physiological levels to ecological and evolutionary considerations (Bale 1987, Baust & Rojas 1985, Block 1990, Cannon & Block 1988, Danks 1987, Denlinger 1991, Duman et al. 1995, Leather et al. 1993, Lee 1989, Lee & Denlinger 1991, Ring & Danks 1994, Sømme 1989, Storey & Storey 1988, Tauber et al. 1986, Zachariassen & Lundheim 1992, as well as Chapter 3 in this volume). Consequently, our chapter will primarily consider those aspects of cold-hardiness specifically related to the regulation of supercooling and ice nucleation.

Supercooling and Ice Nucleation

A pure liquid or solution that remains unfrozen at temperatures below its equilibrium freezing point is said to be supercooled (Angell 1982). In the absence of ice nucleating agents small volumes of water (i.e., on the order of a few microliters) readily supercool, sometimes many degrees below their freezing point. In fact, pure water droplets can approach a limit of $-40°C$ before the random clustering of water molecules spontaneously forms an ice embryo upon which an ice lattice can form, a process termed homogeneous ice nucleation.

In biological systems, ice nucleation almost always occurs at temperatures that are above $-20°C$ (Vali 1995). In this situation it is thought that nucleation occurs via a heterogeneous process in which a non-water substrate functions as the embryonic seed crystal initiating freezing. Relatively inefficient ice nucleators are active at temperatures below $-10°C$, while a few inorganic and organic substances are active at $-5°C$ or warmer. A few bacteria and fungi (discussed later in this chapter) have the unique capacity to catalyze ice nucleation at temperatures near $-2°C$.

The specific subzero temperature at which ice nucleation occurs is determined by a stochastic process that is influenced by both volume and the duration of exposure (Vali 1995). As volume increases, the capacity of a solution to supercool decreases, whereas increasing the duration of exposure to low temperature increases the likelihood that heterogeneous ice nucleation will occur.

Supercooling and Ice Nucleation in Insects

With respect to volume, insects are, in one sense, small bags of water and consequently, in the absence of endogenous ice nucleators, have an inherent capacity to supercool, sometimes extensively (Lee 1989). Many small species and insect eggs supercool by 20 to 30°C before they spontaneously freeze (Sømme 1982). In insects the temperature at which ice nucleation occurs is termed the supercooling point. Experimentally, this value is readily determined by monitoring an insect's body temperature with thermistors or thermocouples as it is cooled to detect the abrupt appearance of an exotherm caused by the release of the latent heat of crystallization as body water freezes. The temperature at which the exotherm begins is the supercooling point.

At the organismal level, the supercooling point is significant for a number of reasons. In the many insects that are unable to survive the freezing of their body water, this value represents the lower lethal temperature. However, some insects are lethally injured when they are cooled to temperatures considerably above their supercooling point (Bale 1987, Lee & Denlinger 1985). Freezing-intolerant insects commonly depress their supercooling points during the autumn in preparation for winter, thereby decreasing the chance that they will freeze internally. Some insects are freezing tolerant and can survive the freezing of 65% or more of their body water (Lee 1991). In contrast to freezing-intolerant species, freezing-tolerant insects often undergo physiological changes that increase their supercooling point during cold-hardening. It is generally believed that promoting internal ice formation at relatively high subzero temperatures functions to slow the rate of extracellular ice formation and consequent cellular dehydration which thereby allows the insect to more easily adjust to this radical change in its internal milieu (Lee 1991).

Many insects can physiologically regulate their supercooling capacity. During cold-hardening (the acquisition of increased cold tolerance) many insects accumulate high concentrations of low molecular mass sugars and polyhydric alcohols, sometimes reaching multimolar levels in the hemolymph (Lee 1991). Glycerol, sorbitol, and trehalose are the most commonly accumulated substances, although others such as fructose, glucose, and mannitol have been reported. One effect of these compounds, sometimes termed low molecular mass antifreezes, is to colligatively depress not only the freezing point, but also the supercooling point. In insects with these antifreezes the supercooling point is depressed by approximately twice as much as the freezing point (Zachariassen 1985). Antifreeze proteins also appear to play a role in promoting supercooling in insects (Duman et al. 1995).

Several sites of ice nucleation and types of endogenous ice nucleators have been identified in insects (Cannon & Block 1988, Lee et al. 1993a, Zachariassen 1992). The gut is the most commonly identified site of ice nucleation. Cessation of feeding or emptying of the gut in preparation for overwintering is often associated with an increased capacity for supercooling. Freezing-tolerant insects commonly produce ice nucleating proteins and lipoproteins that function to limit supercooling and promote freezing at relatively high subzero temperatures (Zachariassen & Hammel 1976). These proteins are efficient ice nucleators inducing freezing at temperatures between -6 to -9°C (Duman et al. 1995). Recently, another class of crystalloid deposits was described that function as heterogeneous nucleators in insects. In larvae of the freezing-tolerant gall fly *Eurosta solidaginis* spherules of calcium phosphate in the Malpighian tubules exhibited ice nucleating activity similar to the temperature at which the intact larvae froze (Mugnano et al. 1996).

Another way in which ice nucleation within the body fluids of an insect may begin is by inoculative freezing (Lee et al. 1996a). In this case, ice external to the insect makes contact with body water and initiates internal freezing. Because this type of freezing may occur with little or no supercooling of body fluids, it has been suggested that the term temperature of crystallization is more universal and appropriate than supercooling point (Wasylyk et al. 1988). Furthermore, inoculative freezing appears to be an important factor for low temperature survival in a number of freezing-tolerant species (Fields & McNeil 1986, Gehrken & Southon 1992, Gehrken et al. 1991) but is deleterious in freezing-intolerant species.

We should emphasize that the supercooling point as determined in the laboratory under idealized conditions necessarily represents the best-case scenario for supercooling capacity. Under field conditions, various factors undoubtedly constrain an individual's potential for supercooling. For example, supercooling capacity of larvae of the goldenrod gall fly changes seasonally in accordance with the amount of moisture within tissues of the gall it inhabits, because this soft-bodied larva is highly susceptible to inoculative freezing (Layne et al. 1990). Early in winter, when moisture is abundant, larvae within galls may freeze at only several degrees below 0°C. In contrast, supercooling point values determined for this species under idealized (i.e., dry) conditions in the laboratory may be as low as - 10°C (Layne et al. 1990). This example underscores the importance of using care in estimating lower lethal temperatures from laboratory supercooling point data (Bale 1987).

Ice Nucleating Active Microorganisms

In the 1970's, ice nucleating active bacteria were discovered in association with plants and decaying leaves (for a historical review see Upper & Vali 1995). Taxonomically, these bacteria are restricted to only a few genera of Gram-negative rods within the Pseudomonadaceae and Enterobacteriaceae. Several recent reviews have summarized molecular and biochemical aspects of bacterial ice nuclei (Fall & Wolber 1995, Kajava

1995, Warren 1995, Wolber 1993, Wolber et al. 1995). The ice nucleating phenotype is due to a minor outer membrane-bound protein whose activity is generally lost during cell fractionation. Both free-living fungi and lichen mycobionts with ice nucleating activity are known (Ashworth & Kieft 1995), however, their highest levels of ice nucleating activity are less than those of bacterial strains. Fungal ice nuclei exhibit greater stability at high temperatures and extremes of pH than bacterial ice nucleators (Pouleur et al. 1992, Fields et al. 1995).

Even if a bacterial strain carries the gene for ice nucleating activity, its phenotypic expression generally varies considerably from cell to cell, even in the same culture (Lindow et al. 1978). Few cells from a given population will exhibit the highest levels of ice nucleating activity at temperatures near $-2°C$, whereas others exhibit considerably less activity. To quantitatively characterize the ice nucleating activity of a bacterial population, Vali (1971, 1995) developed a freezing droplet assay. Various cultural conditions including the composition of the medium and the incubation at low temperature sometimes cause an increase in the expression of the ice nucleating phenotype (Fall & Wolber 1995, Kajava 1995, Warren 1995, Wolber 1993, Wolber et al. 1995).

Because most strains of epiphytic ice nucleating bacteria are not only plant pathogens but are also responsible for extensive frost-related crop losses, they have received considerable study (Hirano & Upper 1991, 1995, Lindow 1983, 1995). When these epiphytic bacteria nucleate water on their own surface they also induce freezing and may facilitate their invasion of their hosts' tissues (Lindow 1983). One novel approach that has considerable promise for controlling these frost-related crop losses uses non-ice nucleating active bacteria to competitively displace or colonize the surface of plants before ice nucleating active bacteria do so (Lindow 1995).

Natural Associations Between Ice Nucleating Active Microorganisms and Insects

For nearly 20 years ice nucleating active microbes were known only from free-living or epiphytic strains. Early in the 1990's, reports appeared describing ice nucleating active microbes that had been isolated from the gut of ectothermic animals, primarily insects (Table 4.1). Strains of *Enterobacter taylorae* and *E. agglomerans* isolated from beetles exhibited

TABLE 4.1 Ice nucleating active bacteria and fungi from the gut of animals

Microorganism	Threshold of Ice Nucleating Activity (°C)	Host Animal	Reference
Bacteria			
Enterobacter agglomerans	–2	*Ceratoma trifurcata* (Coleoptera)	Strong-Gunderson et al. 1990
		Hippodamia convergens (Coleoptera)	Lee et al. 1991
E. taylorae	–2	*C. trifurcata*	Strong-Gunderson et al. 1990
		H. convergens	Lee et al. 1991
Erwinia herbicola	>–10	*Plutella xylostella* (Lepidoptera)	Kaneko et al. 1991a, b
Pseudomonas fluorescens	––	*Dendroides canadensis* (Coleoptera)	Duman et al. 1995
P. putida	–2	*Rana sylvatica* (wood frog)	Lee et al. 1995a
Fungi			
Fusarium sp.	–5	*Chilo suppressalis* (Lepidoptera)	Tsumuki et al. 1992

maximal thresholds of ice nucleating activity at approximately -2°C, only slightly less than the highly active epiphytic strains (Lee et al. 1991). When these bacteria were fed to an insect model, the lady beetle *Hippodamia convergens*, its supercooling point increased by 12-13°C above its unfed control level of -16°C (Lee et al. 1991). Similarly, Kaneko and colleagues (1991a,b) isolated *Erwinia herbicola*, which had significant ice nucleating activity, from the diamondback moth, *Plutella xylostella*.

Recent investigations with the rice stem borer, *Chilo suppressalis*, described an ice nucleating active fungus isolated from its gut flora that had sufficient ice nucleating activity to explain fully the supercooling point (-8.4°C) of the intact larvae (Tsumuki 1992, Tsumuki & Konno 1991). Unlike the cases of ice nucleating bacterial strains that were isolated from freezing-intolerant insects, the rice stem borer is freezing tolerant. Consequently, this result indicates that the fungal ice nucleator functions like ice nucleating proteins to insure that protective freezing will begin at high subzero temperatures and suggests a mutualistic association between the fungus and its insect host (Lee et al. 1993b, 1995b, Tsumuki 1992). Of particular note, an ice nucleating active *Pseudomonas putida* has been isolated from the gut of the wood frog, *Rana sylvatica*, and may, under certain conditions, serve a similar function for this freezing tolerant species (Costanzo & Lee 1996, Lee et al. 1995a).

Manipulation of Insect Supercooling Using Ice Nucleating Active Microorganisms

It is now evident that the supercooling capacity of a wide variety of insects is readily manipulated using ice nucleating active microorganisms (Lee et al. 1993a, 1995b). Either living or killed preparations of these ice nucleators have been used to significantly increase the supercooling point of adults and/or larvae of five insect orders: Coleoptera, Diptera, Hemiptera, Hymenoptera, and Lepidoptera (Table 4.2). Depending on the species, these treatments increase the supercooling point by a few degrees to more than 15 degrees Celsius. Ingestion of ice nucleating active bacteria causes an immediate elevation of the supercooling point (Strong-Gunderson et al. 1990); after an insect drinks for only a few seconds from a solution of ice nucleating bacteria we have observed an elevation within the few minutes required to make a supercooling point determination.

TABLE 4.2 Effect of treatment with ice nucleating active microorganisms on the supercooling point of insects. (Expanded and modified from Lee et al. 1993a)

Insect (Stage)	Microorganism	Supercooling Point (°C) (± SEM)		Reference
		Untreated	Treated	
Coleoptera				
Cryptolestes ferrugineus (adult)	Pseudomonas syringae[5]	-17.0 ± 1.0	-8.1 ± 0.5	Fields 1990
C. pusillus (adult)	P. syringae[5]	-14.0 ± 1.0	-12.0 ± 1.5	Fields 1992
Diabrotica undecimpunctata howardi (adult)	P. syringae[1]	-7.5 ± 0.8	-3.2 ± 0.2	Strong-Gunderson et al. unpub. data
Gibbium psylloides (adult)	P. syringae[3]	-10.7 ± 0.9	-6.0 ± 0.5	Lee et al. 1992b
Hippodamia convergens (adult)	P. syringae[1]	-16.0 ± 0.5	-2.8 ± 0.2	Strong-Gunderson et al. 1990
	Erwinia herbicola[4]	-16.0 ± 0.5	-4.4 ± 0.6	Strong-Gunderson et al. 1990
	Enterobacter agglomerans[4]	-16.0 ± 0.5	-3.1 ± 0.1	Lee et al. 1991
	E. taylorae[4]	-16.0 ± 0.5	-4.3 ± 0.4	Lee et al. 1991
	Fusarium acuminatum[7]	-14.9 ± 0.5	-11.0 ± 0.7	Lee et al. 1992b
Oryzaephilus surinamensis (adult)	P. syringae[5]	-13.7 ± 1.9	-11.0 ± 1.3	Fields 1992
Rhyzopertha dominica (adult)	P. syringae[3]	-15.2 ± 0.6	-3.3 ± 0.1	Lee et al. 1992b
Sitophilus granarius (adult)	P. syringae[5]	-14.3 ± 0.8	-7.8 ± 0.5	Fields 1992
S. granarius (adult)	P. syringae[3]	-15.7 ± 1.0	-8.0 ± 0.6	Lee et al. 1992b
Tenebrio molitor (larva)	P. syringae[2]	-16.0 ± 0.7	-5.4 ± 0.7	Strong-Gunderson et al. unpub. data
T. molitor (adult)	P. syringae[2]	-15.1 ± 0.6	-2.7 ± 0.3	Strong-Gunderson et al. unpub. data

Tribolium castaneum (adult)	*P. syringae*[3]	-13.9 ± 0.8	-4.7 ± 0.4	Lee et al. 1992b
	P. syringae[5]	-12.3 ± 1.0	-5.8 ± 0.3	Fields 1992
Diptera				
Sarcophaga crassipalpis (larva)	*P. syringae*[2]	-13.8 ± 0.9	-3.6 ± 0.1	Strong-Gunderson et al. unpub. data
Hemiptera				
Lygus sp. (adult)	*P. syringae*[1]	-20.0 ± 0.5	-8.7 ± 1.0	Strong-Gunderson & Lee, unpub. data
Hymenoptera				
Solenopsis invicta (adult)	*P. syringae*	-7.9 ± 0.6	-4.1 ± 0.9	Landry & Phillips 1996
Lepidoptera				
Chilo suppressalis (larva)	*Fusarium* sp.[6]	-20.1 ± 0.9	-5.7 ± 0.6	Tsumuki 1992
Galleria mellonella (larva)	*P. syringae*[2]	-10.4 ± 0.1	-4.0 ± 0.3	Strong-Gunderson et al. unpub. data
Plodia interpunctella (larva)	*P. syringae*[3]	-10.3 ± 0.4	-5.4 ± 0.5	Lee et al. 1992b

[1] Misted with 10^9 bacteria/ml water.
[2] Misted with 10^8 bacteria/ml water.
[3] Treated with 100 ppm dry, powdered bacteria.
[4] Ingestion of 2×10^9 bacteria/ml water.
[5] Treated with 1,000 ppm dry, powdered bacteria.
[6] Ingestion.
[7] Misted with 3 mg/ml water.

Of particular interest is the fact that the supercooling point of insects, whose mouths have been sealed to prevent ingestion, is readily elevated by applying various preparations of living and dead ice nucleating active microorganisms to the cuticle (Strong-Gunderson et al. 1992). Steigerwald et al. (1995) recently explored several non-oral avenues by which surface application of nucleating agents might make contact with, and initiate the freezing of, the body water of insects. Using cold-hardy adults of the beetle *H. convergens* which consistently maintain low supercooling points of approximately $-16°C$, a suspension of *P. syringae* was applied to four anatomical sites (Fig. 4.1). Compared with control treatments, aqueous suspensions of either cultured or killed *P. syringae* suspensions produced significantly higher mean values when applied to the thoracic spiracle of the insect, $-7.7°$ and $-5.6°C$, respectively. Similarly, application of the ice nucleating active fungus *Fusarium avenaceum* to the thoracic spiracle significantly elevated the supercooling point from $-16°C$ to approximately $-10°C$. Consequently, the spiracles may provide direct access to the body water of insects and explain, at least in part, the relative ease with which the supercooling capacity of insects is diminished using surface application of ice nucleating microbes.

Potential Use of Ice Nucleating Active
Microorganisms for Biological Control

The fact that the supercooling point of a wide variety of insects can be elevated using ice nucleating active microorganisms supports the proposition that these ice nucleators may be used for the biological control of insect pests (see reviews by Fields 1992, Lee 1991, Lee et al. 1993a, 1995b). Because most of such species are intolerant of internal freezing, these microbial ice nucleators could be used to reduce these insects' capacity to supercool and thereby compromise their ability to survive the low temperatures of winter. Obviously this strategy for control is only feasible if insects naturally experience temperatures below this elevated supercooling point in their overwintering site. Another significant problem to be overcome is how to deliver the ice nucleating active microorganism to the insect pest and have it retain its activity until low temperatures are experienced. Nevertheless, the use of ice nucleating active microbes for pest control has the advantages of avoiding toxic chemicals or the release of

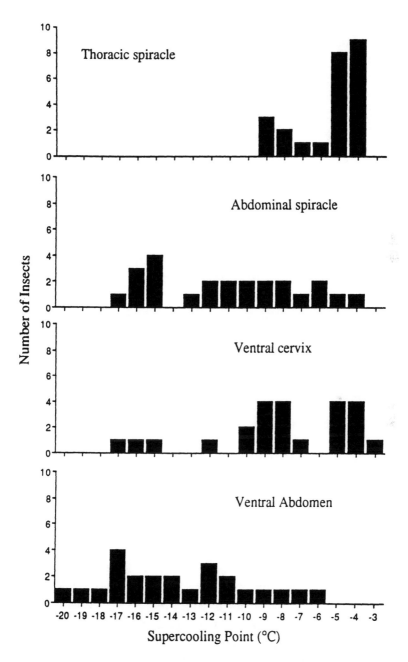

FIGURE 4.1 Distribution of supercooling points of the freezing intolerant beetle, *Hippodamia convergens.* Inoculum volume was 0.5 μl of 20,000 ppm UVI *Pseudomonas syringae.* Site of inoculation listed with corresponding graph (from Steigerwald et al. 1995).

genetically altered miroorganisms into the field, and it is biodegradable and compatible with other forms of pest management (Lee et al. 1993a).

In this decade several research groups have worked on problems directly related to the potential use of this approach for biological control (Fields 1992, Hong et al. 1994, Landry & Phillips 1996, Lee et al. 1991, Strong-Gunderson et al. 1990). Several studies have focused on controlling pests of stored products, particularly those infesting granaries, by exploiting the fact that the supercooling point of a variety of these insects is readily elevated using ice nucleating active bacteria and fungi (Fields 1990, 1993, Fields et al. 1995, Lee et al. 1992b). However, in some geographic locations the temperature within grain storage bins may not normally fall low enough to induce internal freezing of the insects even in the presence of biological ice nucleators, or the electrical costs of cooling the grain may be prohibitive (Fields 1993). Because these studies have been reviewed recently elsewhere (Fields 1992, Lee et al. 1993a, 1995b), the remainder of this chapter will focus on our recent efforts to answer basic and applied questions related to the efficacy of killed and living ice nucleating active microbes in elevating the supercooling point, modes of delivery of these agents to the pest insects, and the significance of microclimatic conditions within the hibernaculum, using the Colorado potato beetle, *Leptinotarsa decemlineata*, as an insect model system. Unlike pests infesting stored products located in relatively controlled environments, this beetle is representative of species that naturally experience subzero temperatures in natural habitats. If ice nucleating active microbes are to be used for control of these species, they must be able to function under a variety of environmental conditions.

Regulation of Ice Nucleation in the Colorado Potato Beetle

Well-known for rapidly developing resistance against a wide range of pesticides, including synthetic pyrethroids, the Colorado potato beetle is the most serious pest of potatoes in North America (Casagrande 1987, Ioannidis et al. 1991). The current agricultural practice of planting extensive monocultures of potatoes further exacerbates the problem of progressive population growth of the beetles from year to year (Casagrande 1987). Consequently, alternative methods are needed urgently for the control of this pest.

In late summer or early autumn adult beetles enter shallow burrows in the soil where they overwinter (Mail & Salt 1933, Ushatinskaya 1978). Burrow soil temperature and moisture influence overwintering survival (Fink 1925, Lashomb et al. 1984, Weber & Ferro 1993). Recent attempts to devise cultural control methods used trap crops on the edges of fields late in the summer as a means of concentrating adults in restricted areas (Kung et al. 1992, Milner et al. 1992). These areas were covered with an insulating layer of mulch in an attempt to limit the depth to which the beetles would burrow as they prepared to overwinter. It was hypothesized that removal of mulch, and therefore the insulation it provided, in mid-winter would cause a rapid decrease in soil temperature and kill the beetles which had remained in superficial burrows. In fact, Milner et al. (1992) reported greater mortality in sites where the mulch was removed relative to control plots.

Our initial idea was to complement this cultural approach by using ice nucleating active microbes to further increase the susceptibility of the beetles to low temperature. In our first attempt to elevate the supercooling point of the Colorado potato beetle we exposed beetles to a concentrated, freeze-dried, and killed preparation of *P. syringae* (Genencor International, Rochester, NY) mixed with soil (Lee et al. 1994). Untreated beetles had mean supercooling points of $-7.6 \pm 0.2°C$ (Fig. 4.2A). During both years of the study, treatment with 1 to 1,000 ppm of the *P. syringae* preparation elevated supercooling points in a dose-dependent manner as reported for other insects (Fields 1990, Lee et al. 1992a). The highest values of $-3.7 \pm 0.1°C$ resulted from treatment with 1,000 ppm; however, application with as little as 1 ppm resulted in a significant increase in the supercooling point as compared to untreated control beetles. When these data were plotted as the cumulative percentage of beetles frozen versus the exposure temperature (Fig. 4.2B), it clearly showed the population range of supercooling point values following a given treatment. It also allowed us to predict the proportion of the beetles expected to survive exposure to a given subzero temperature. For beetles treated with 10 ppm, approximately 75% would be expected to freeze by the time the environmental temperature was lowered to $-7°C$ (Fig. 4.2B).

The fact that field-collected adults have a relatively high supercooling point $(-7°C)$ and do not survive prolonged freezing at this temperature indicates that this species has rather limited cold-hardiness that is consistent with their thermally buffered hibernaculum within the soil (Lee et al. 1995b). Nonetheless, these data suggest that an elevation of as little as

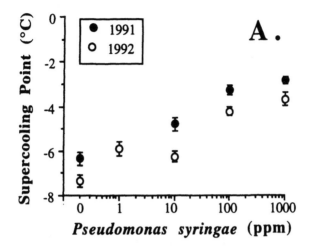

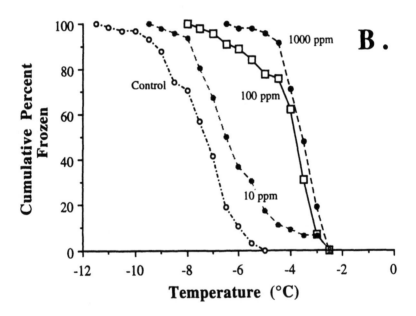

FIGURE 4.2 (A) Effect of *Pseudomonas syringae* on the mean (±SEM) supercooling point of overwintering adults of the Colorado potato beetle. Beetles were exposed to various concentrations (0 to 1,000 ppm) of *P. syringae* in soil for 48 h at 4°C. In 1991 sample sizes were n = 10-11; in 1992, n = 44-58. (B) cumulative freezing profile for beetles exposed to various concentrations of *P. syringae* in 1992 (from Lee et al. 1994).

2-3 degrees Celsius in the supercooling point, caused by ice nucleating active microbes, would cause a significant decrease in the chance that they would survive the winter.

Site of Bacterial Application Affects Supercooling Point

Using an approach similar to that of Steigerwald et al. (1995), we investigated the effect of topical application of *P. syringae* on the supercooling capacity of the Colorado potato beetle (Lee et al. 1996b). Application of *P. syringae* to the ventral abdomen did not significantly increase the supercooling point (-5.5 °C) compared with beetles treated with the non-ice nucleating active (control) bacterium *Escherichia coli* (Table 4.3). In contrast, application of *P. syringae* to the thoracic spiracle, ventral cervix, or abdominal spiracle significantly elevated supercooling point values. Taken together with the data from *H. convergens* (Fig. 4.1), these results indicate that application of ice nucleating active microbes to a number of non-oral sites can be used to elevate the supercooling point of insects. These data also suggest that it may be relatively difficult, compared to the development of resistance to traditional chemical insecticides, for an insect to develop resistance to the action of these agents for biological control.

TABLE 4.3 Supercooling point values after application of an aqueous suspension of either a non-ice nucleating active bacterial control *Escherichia coli* or the ice nucleating active *Pseudomonas syringae* to four anatomic sites on the Colorado potato beetle. Values identified by different letters are statistically distinguishable (Lee et al. 1996b)

Treatment	Anatomic Site of Application	n	Supercooling Point (°C) (Mean ± SEM)
E. coli	Thoracic spiracle	28	-6.5 ± 0.2a
P. syringae	Ventral abdomen	32	-5.5 ± 0.2ab
	Abdominal spiricle	32	-4.7 ± 0.2bc
	Ventral cervix	20	-5.1 ± 0.3bc
	Thoracic spiracle	26	-4.5 ± 0.3c

Surfactant Enhances Ice Nucleating Active Fungus

Recent studies in our laboratory demonstrated that surface application of the filamentous ice nucleating active fungus *Fusarium acuminatum* elevates the supercooling point of *H. convergens* (Lee et al. 1992a). Using an aqueous suspension of *F. acuminatum* (0.03 g/ml) in the freeze-drop assay, 50% of 10 µl drops froze at -6.1°C or higher. When beetles were misted with this suspension, their supercooling points increased slightly from -14.9°C (misted with water only) to -11.0°C. In an attempt to further increase the supercooling point, we added surfactants to the fungal suspension under the assumption that greater contact between surface water and the cuticle might be achieved if the surface tension was reduced. Notably, when fungi suspended in a 1% solution of the surfactant Tween 80 were applied, the supercooling point increased from -14.9° to -5.8°C. These results suggest that surfactants used in combination with ice nucleating active microbes may be useful in the development of protocols for the control of insect pests.

Effect of Soil Moisture and Composition on Colorado Potato Beetle Cold Hardiness

To use ice nucleating active microbes under field conditions it is necessary to have a thorough understanding of the natural mechanisms of cold-hardiness of the insect within its natural hibernaculum. For the Colorado potato beetle, soil moisture appears to play an important role in its overwintering biology. Although Tauber et al. (1994) indicated the importance of soil moisture levels in regulating dormancy and subsequent emergence of Colorado potato beetles in spring, little is known concerning the interaction between substrate moisture and cold hardiness in insects that overwinter within the soil.

Consequently, we recently investigated the role of soil moisture and hydric variables in the winter cold hardiness of the Colorado potato beetle (Costanzo et al. 1997). Diapausing adults chronically exposed to sandy soil exhibited body mass and body water content changes that were dependent on soil moisture content. These changes in body water content, in turn, influenced the supercooling point (Fig. 4.3; range, -3.3° to -18.4°C), indicating that environmental moisture indirectly determined supercooling capacity. Tests involving acute chilling of beetles showed that specimens chilled in dry sand readily tolerated a 24-h exposure to tempera-

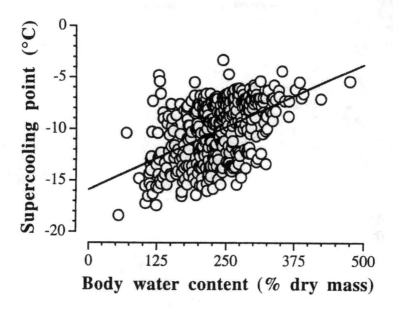

FIGURE 4.3 Correlation of body water content (% dry mass) and supercooling point of Colorado potato beetles during diapause at 4°C, DD (from Costanzo et al. 1997).

tures as low as −5°C, but beetles tested in even slightly damp sand (e.g., water content: 1.7% of dry mass) incur high mortality (Fig. 4.4). Apparently, burrowing in dry soils not only promotes supercooling via its effect on water balance, but also inhibits inoculative freezing of Colorado potato beetles.

Costanzo et al. (1997) also reported that mortality of beetles was strongly influenced by substrate texture, because survival of beetles exposed to −5°C for 24 h was higher in substrates composed of sand, clay, and/or peat (48-64%) than in pure silica sand (22%). They concluded that not only moisture, but also texture, structure, water potential, and related physico-chemical attributes of soil may strongly influence the cold hardiness and overwintering survival of burrowing insects. Furthermore, this work indicated that manipulating the moisture levels of the soil surrounding Colorado potato beetles during winter may compliment the action of ice nucleating active microbes applied to this species.

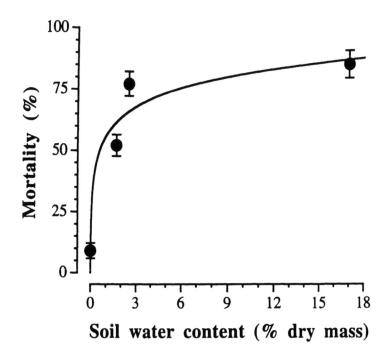

FIGURE 4.4 Mortality of Colorado potato beetle chilled at −5°C for 24 h in sand containing various amounts of moisture. Mean values (± 1 SE) are based on n = 200. Data are fitted with logarithmic curve (from Costanzo et al. 1997).

Longevity of Ice Nucleating Active Pseudomonas syringae *Preparation in the Soil*

In a another series of experiments we examined the length of time that a freeze-dried, killed preparation of *P. syringae* could retain its ice nucleating activity in soil under simulated field conditions. To test the effect of temperature and soil moisture on the ice nucleating activity, a 100 ppm *P. syringae* preparation was added to soil and held for 16 weeks at 4 or 15°C. Periodically, diapausing Colorado potato beetle adults were added to the soil for 1-2 hours, removed and their supercooling points determined. Overall ice nucleating activity was retained better in dry, as compared to moist, soil and at cooler versus warmer conditions, results that are consistent with a previous study that reported a loss of activity for this material at relatively high temperatures (Goodnow et al. 1990).

Seasonal Characterization of the Normal
Gut Flora in the Colorado Potato Beetle

As an alternative delivery method to the addition of ice nucleating active microbes to the surface of the beetles before they burrow into the soil or to adding them to the soil directly, we are evaluating the feasibility of colonizing the beetle's gut with living bacteria or fungi. Our first step, which is nearly complete, in this line of investigation was to characterize seasonal gut flora in the Colorado potato beetle. In addition to identifying the bacterial flora, we also screened them for ice nucleating activity. Adult beetles collected from potato plants in the summer revealed a predominance of *Enterobacter taylorae* and *E. agglomerans* strains. Low levels of ice nucleating activity were detected in multiple strains of both microbes from the gut of the Colorado potato beetle. These data are consistent with previous reports that the ice nucleating active phenotype in bacteria isolated from insects has only been found in Gram-negative, aerobic rods in the Pseudomonadaceae and Enterobacteriaceae (Lee et al. 1991, 1993a). Less abundant bacterial species present as normal flora included *Serratia marcescens, Klebsiella pneumoniae, Klebsiella oxytoca,* and *Xanthomonas maltophilia*; ice nucleating activity was not detected in these strains. The common ice nucleating active epiphyte *P. syringae* was not found. Of particular interest was the fact that even field-collected over-wintering beetles that did not feed for at least three months, and whose gut appeared shrunken and relatively or completely empty, retained a gut flora similar to that found in summer adults. Although the bacterial populations were apparently reduced compared to summer-collected adults, they did retain a similar diversity of normal flora through the winter. This result supports our idea to establish and maintain ice nucleating active microbes in the gut flora through the winter.

Isolation and Characterization of Pseudomonas putida

In related studies our research group isolated ice nucleating active bacteria from the gut of winter-collected, freezing-tolerant wood frogs (Lee et al. 1995a). Multiple strains of *P. fluorescens, P. putida,* and *E. agglomerans* with ice nucleating activity were identified. The *P. putida* strains exhibited substantial levels of ice nucleation activity ranging from -1.6° to -3.0°C, which places them among the most potent of known microbial nucleators. This activity was confirmed *in vivo* by feeding them to another freeze-tolerant frog *Pseudacris crucifer* resulting in a decreased capacity for this frog to supercool and remain unfrozen at -2°C (Lee et al. 1995a). Similar to the report by Tsumuki (1992) these bacteria may play a role in enhancing winter survival by promoting ice nucleation at high subzero temperatures. This research is germane to this project because we have isolated ice nucleating active *P. putida* from an insect previously, this strain has high levels of ice nucleating activity, and, as described in the next section, ingestion of one of these *P. putida* strains by the Colorado potato beetle caused a significant elevation of the supercooling point for at least 2.5 months (Table 4.4).

TABLE 4.4 Supercooling points of Colorado potato beetles collected in mid-September 1994 from cultivated potato fields in central Wisconsin fed various bacterial suspensions on potato slices (J. P. Costanzo, T. L. Reed, R. E. Lee, J. B. Moore, and M. R. Lee, unpublished data)

| Feeding Treatment | Supercooling Point (°C, mean ± SEM, n = 20) at Various Intervals after Feeding | | |
	<1.5 h	14 days	2.5 months
Unfed control	-6.8 ± 0.1	--	--
Escherichia coli (non ice-nucleating control)	-6.9 ± 0.1	-7.2 ± 0.2	-6.4 ± 0.2
Enterobacter agglomerans	-4.7 ± 0.1*	-6.3 ± 0.3	-6.8 ± 0.4
Pseudomonas syringae			
live	-4.1 ± 0.2*	-6.9 ± 0.1	-5.8 ± 0.3*
freeze-dried killed	-4.4 ± 0.1*	-7.4 ± 0.3	-6.9 ± 0.3
P. flourescens	-4.2 ± 0.1*	-4.3 ± 0.3*	-3.9 ± 0.3*
P. putida	-3.9 ± 0.1*	-5.0 ± 0.4*	-3.4 ± 0.1*

* mean differs significantly from mean for unfed group (Fisher's PLSD; $P < 0.05$)

Persistence and Activity of Ice Nucleating
Active Bacteria in the Colorado Potato Beetle

One of the challenges that must be met if ice nucleating active microorganisms are to be used for pest control is to find ways to deliver these microorganisms to the insect pests. Consequently, we tested the efficacy of several different species of living ice nucleating active bacteria and fungi for their effect on the Colorado potato beetle's supercooling point. Suspensions of living *P. fluorescens, P. syringae,* and *P. putida* sprayed onto adults all caused a significant increase in the supercooling point, indicating that living bacterial cells also may be used for supercooling point manipulation.

Recently we completed a preliminary study demonstrating that ingestion of living ice nucleating active bacteria caused an elevation of the beetle's supercooling point for at least ten weeks under conditions that simulated natural overwintering. Beetles were collected in early autumn and fed on various ice nucleating active bacteria added to potato tuber before they entered diapause. Beetles readily fed on a slice of tuber that was coated with the bacterial solution. Beetles were assayed for super-cooling capacity and gut flora within 1.5 h after feeding, at the conclusion of a 14-day diapause induction regimen, and in mid-winter, after the beetles had been in diapause 2.5 months.

As expected, ingestion of ice nucleating active bacteria immediately caused an increase in beetle supercooling points (Table 4.4). Most noteworthy was the fact that ingestion of *P. putida* derived from the freeze-tolerant frog and *P. fluorescens* caused not only an initial supercooling point elevation, but the supercooling point remained elevated for the entire 2.5 months of this study. It is also significant that the supercooling point remained elevated even though the beetles purged their guts extensively in preparation for burrowing beneath the soil and overwintering in reproductive diapause. These results confirm and extend findings of Chapco & Kelln (1994) in which ingested bacteria were retained in the gut of grasshoppers for more than three weeks. Furthermore, these data lend credence to the idea of establishing ice nucleating active bacteria in the insect gut as a strategy for delivering ice nucleating active microorganisms to insect pests.

Future Directions

These studies represent only the first steps toward improving our understanding of both the natural flora of insects as well as the potential for establishing ice nucleating active microbes in insect pests. The establishment of microbes in insect pests has a wide range of possible applications for biological control. For example, the use of self-generating entomopathogenic bacteria is a likely control strategy (Chapco & Kelln 1994). Because our long-term goal is to use these bacteria for biological control, it is critical that we know the identity and ice nucleating activity of naturally occurring ice nucleating active microorganisms and their potential for manipulating their association with insect pests.

Our progress thus far provides a solid foundation for field tests with overwintering Colorado potato beetles that will attempt to elevate their supercooling point by augmenting soil moisture levels and by using a freeze-dried, killed and/or live preparations of *P. syringae*. In a complementary line of investigation we will continue to study the use of living ice nucleating active microorganisms to colonize the insect gut or insect surface so that they function to diminish the overwintering beetle's supercooling capacity and promote death by freezing. It may be possible to not only colonize the gut but to develop strains of microbes with higher levels of ice nucleating activity, or ones whose expression of the ice nucleating active phenotype increases as environmental temperatures decrease during the autumn.

If practical protocols can be developed for manipulating insect cold hardiness, this approach has the potential for controlling a variety of insect pests. Pests of stored products in which it is possible to manipulate the environmental temperature in conjunction with microbial treatments may be particularly good candidates for this type of control. Futhermore, the results of laboratory investigations with ice nucleating active microbes thus far suggest that methods can be developed for control of selected insect pests in agricutural environments.

Acknowledgments

We thank Jackie Litzgus for her comments on the manuscript. This research was supported by the Cooperative State Research Service (USDA) grant #96-35302-3419, NSF grant IBN-9305809, and Genencor Intl., San Francisco.

References

Angell, C. A. 1982. Supercooled Water, *In* F. Franks, ed. *Water-A Comprehensive Treatise*, pp 1-82. Plenum Press, New York.

Ashworth, E. N. & Kieft, T. L. 1995. Ice nucleation activity associated with plants and fungi, *In* R. E. Lee, G. J. Warren & L. V. Gusta, eds. *Biological Ice Nucleation and Its Applications.* pp 137-162. American Phytopathological Society, St. Paul, Minnesota.

Bale, J. S. 1987. Insect cold hardiness: Freezing and supercooling-an ecophysiological perspective. J. Insect Physiol. 33: 899-908.

Baust, J. G. & R. R. Rojas. 1985. Review - insect cold hardiness: facts and fancy. J. Insect Physiol. 31: 755-759.

Block, W. 1990. Cold tolerance of insects and other arthropods. Phil. Trans. R. Soc. Lond. B. 326: 613-633.

Cannon, R. J. C. & W. Block. 1988. Cold tolerance of microarthropods. Biol. Rev. 63: 23-77.

Casagrande, R. A. 1987. The Colorado Potato Beetle: 125 years of mismanagement. Bull. Ent. Soc. Am. 142-150.

Chapco, W. & R. A. Kelln. 1994. Persistance of ingested bacteria in the grasshopper gut. J. Invert. Pathol. 64: 149-150.

Costanzo, J. P. & R. E. Lee. 1996. Mini-review: Ice nucleation in freeze-tolerant vertebrates. Cryo-Letters 17: 111-118.

Costanzo, J. P., J. B. Moore, R. E. Lee, P. E. Kaufman & J. A. Wyman. 1997. Influence of soil hydric parameters on the winter cold hardiness of a burrowing beetle, *Leptinotarsa decemlineata* (Say). J. Comp. Physiol. B. 167: 169-176.

Danks, H. V. 1987. *Insect Dormancy: An Ecological Perspective.* Ottawa: Biological Survey of Canada (Terrestrial Arthropods). 439 pp.

Denlinger, D. L. 1991. Relationship between cold hardiness and diapause, *In* R. E. Lee & D. L. Denlinger, eds. *Insects at Low Temperature.* pp 174-198. Chapman and Hall, New York.

Duman, J. G., T. M. Olsen, K. L. Yeung & F. Jerva. 1995. The roles of ice nucleators in cold tolerant invertebrates, *In* R. E. Lee, G. J. Warren & L. V. Gusta, eds. *Biological Ice Nucleation and Its Applications.* pp 201-219. American Phytopath. Soc., St. Paul, Minnesota.

Fall, R. & P. K. Wolber. 1995. Biochemistry of bacterial ice nuclei, *In* R. E. Lee, G. J. Warren & L. V. Gusta, eds. *Biological Ice Nucleation and Its Applications.* pp 63-83. American Phytopath. Soc., St. Paul, Minnesota.

Fields, P. G. 1990. The cold-hardiness of *Cryptolestes ferrugineus* and the use of ice nucleation-active bacteria as a cold-synergist, *In* F. Fleurat-Lessard and P. Duncon, eds. *Proceedings of the Fifth International Working Conference on Stored-product Protection, Vol. 2.* pp 1183-1191. Bordeaux, France.

Fields, P. G. 1992. The control of stored-product insects and mites with extreme temperatures. J. Stored Prod. Res. 28: 89-118.

Fields, P. G. 1993. Reduction of cold tolerance of stored-product insects by ice-nucleating-active bacteria. Environ. Entomol. 22: 470-476.

Fields, P. G. & J. N. McNeil. 1986. Possible dual cold-hardiness strategies in *Cisseps fulvicollis* (Lepidoptera: Arctiidae). Canad. Entomol. 118: 1309-1311.

Fields, P. G., S. Pouleur & C. Richard. 1995. The effect of high temperature storage on the capacity of an ice-nucleating-active bacterium and fungus to reduce insect cold-tolerance. Canad. Entomol. 127: 33-40.

Fink, D. E. 1925. Physiological studies on hibernation in the potato beetle, *Leptinotarsa decemlineata* Say. Biol. Bull. 6: 381-406.

Gehrken, U. & T. E. Southon. 1992. Supercooling in a freeze-tolerant cranefly larva, *Tipula* sp. J. Insect Physiol. 38: 131-137.

Gehrken, U., A. Strømme, R. Lundheim & K. E. Zachariassen. 1991. Inoculative freezing in overwintering tenebrionid beetle, *Bolitophagus reticulatus* Panz. J. Insect Physiol. 37: 683-687.

Goodnow, R. A., M. D. Harrison, J. D. Morris, K. B. Sweeting & R. J. LaDuca. 1990. Fate of ice nucleation-active *Pseudomonas syringae* strains in alpine soils and waters and in synthetic snow samples. Appl. Environ. Microbiol. 56: 2223-2227.

Hirano, S. S. & C. D. Upper. 1991. Bacterial community dynamics. *In* J. H. Andrews & S. S. Hirano, eds. *Microbial Ecology of Leaves.* pp 271-294. Springer-Verlag, New York.

Hirano, S. S. & C. D. Upper. 1995. Ecology of ice nucleation-active bacteria. *In* R. E. Lee, G. J. Warren & L. V. Gusta, eds. *Biological Ice Nucleation and Its Applications.* pp 41-61. American Phytopath. Soc., St. Paul, Minnesota.

Hong, Z., S. Fuzai, Z. Yongxiang & W. Wenlai. 1994. The effect of INA bacteria on the freezing temperature of cotton bollworm. Scientia Agric. Sinica 27: 27.

Ioannidis, P. I., E. J. Grafius & M. E. Whalen. 1991. Patterns of insecticide resistance to azinphosmethly, carbofuran, and permethrin in the Colorado potato beetle (Coleoptera: Chrysomelidae). J. Econ. Entomol. 84: 1417-1423.

Kajava, A. 1995. Molecular modelling of the three-dimensional structure of bacterial Ina proteins, *In* R. E. Lee, G. J. Warren & L. V. Gusta, eds. *Biological Ice Nucleation and Its Applications.* pp 101-114. American Phytopath. Soc., St. Paul, Minnesota.

Kaneko, J., K. Kita & K. Tanno. 1991a. INA bacteria isolated from diamondback moth *Plutella xylostella* L. pupae. Japanese J. Appl. Entomol. Zool. 35: 7-12 (in Japanese).

Kaneko, J., T. Yoshida, T. Owada, K. Kita & K. Tanno. 1991b. *Erwinia herbicola* ice nucleation active bacteria isolated from diamondback moth *Plutella xylostella* L. pupae. Japanese J. Appl. Entomol. Zool. 35: 247-251 (in Japanese).

Kung, K.-J. S., M. Milner, J. A. Wyman, J. Feldman & E. Nordheim. 1992. Survival of Colorado potato beetle (Coleoptera: Chrysomelidae) after exposure to subzero thermal shocks during diapause. J. Econ. Entomol. 85: 1695-1700.

Landry, C. E. & S. A. Phillips. 1996. Potential of ice-nucleating active bacteria for management of the red imported fire ant (Hymenoptera: Formicidae). Environ. Entomol. 25: 859-866.

Lashomb, J. H., Y.-S. Ng, G. Ghidiu & E. Green. 1984. Description of spring emergence by the Colorado potato beetle, *Leptinotarsa decemlineata* (Say) (Coleoptera: Chrysomelidae), in New Jersey. Environ. Entomol. 13: 907-910.

Layne, J. R., R. E. Lee & J. L. Huang. 1990. Inoculation triggers freezing at high subzero temperatures in a freeze-tolerant frog (*Rana sylvatica*) and insect (*Eurosta solidaginis*). Can. J. Zool. 68: 506-510.

Leather, S. E., K. F. A. Walters & J. S. Bale. 1993. *The Ecology of Insect Overwintering*, Cambridge University Press, New York. 255 pp.

Lee, M. R., R. E. Lee, J. M. Strong-Gunderson & S. R. Minges. 1992a. Treatment with ice nucleating active fungi and surfactants decrease insect supercooling capacity. Cryobiology 29: 743.

Lee, M. R., R. E. Lee, J. M. Strong-Gunderson & S. R. Minges. 1995a. Isolation of ice nucleating active bacteria from the freeze-tolerant frog, *Rana sylvatica*. Cryobiology 32: 358-365.

Lee, R. E. 1989. Insect cold-hardiness: to freeze or not to freeze? Bioscience 39: 308-313.

Lee, R. E. 1991. Principles of insect low temperature tolerance, *In* R. E. Lee & D. L. Denlinger, eds. *Insects at Low Temperature*. pp 17-46. New York: Chapman and Hall.

Lee, R. E., J. P. Costanzo, P. E. Kaufman, M. R. Lee & J. D. Wyman. 1994. Ice-nucleating active bacteria reduce the cold-hardiness of the freeze-intolerant Colorado potato beetle (Coleoptera: Chrysomelidae). J. Econ. Entomol. 87: 377-381.

Lee, R. E., J. P. Costanzo & J. A. Mugnano. 1996a. Regulation of supercooling and ice nucleation in insects. Eur. J. Entomol. 93: 405-418.

Lee, R. E. & D. L. Denlinger. 1985. Cold tolerance in diapausing and non-diapausing stages of the flesh fly, *Sarcophaga crassipalpis*. Physiol. Entomol. 10: 309-315.

Lee, R. E. & D. L. Denlinger, eds. 1991. *Insects at Low Temperature*. Chapman and Hall, New York. 513 pp.

Lee, R. E., M. R. Lee & J. M. Strong-Gunderson. 1993a. Insect cold-hardiness and ice nucleating active microorganisms including their potential use for biological control. J. Insect Physiol. 39: 1-12.

Lee, R. E., M. R. Lee & J. M. Strong-Gunderson. 1995b. Biological control of insect pests using ice-nucleating microorganisms, *In* R. E. Lee, G. J. Warren & L. V. Gusta, eds. *Biological Ice Nucleation and Its Applications*. pp 257-269. American Phytopathol. Soc., St. Paul, Minnesota.

Lee, R. E., J. J. McGrath, R. T. Morason & R. M. Taddeo. 1993b. Survival of intracellular freezing, lipid coalescence and osmotic fragility in fat body cells of the freeze-tolerant gall fly *Eurosta solidaginis*. J. Insect Physiol. 39: 445-450.

Lee, R. E., K. A. Steigerwald, J. A. Wyman, J. P. Costanzo & M. R. Lee. 1996b. Anatomic site of application of ice-nucleating active bacteria affects supercooling in the Colorado potato beetle (Coleoptera: Chrysomelidae). Environ. Entomol. 25: 465-469.

Lee, R. E., J. M. Strong-Gunderson, M. R. Lee & E. C. Davidson. 1992b. Ice-nucleating active bacteria decrease the cold-hardiness of stored grain insects. J. Econ. Entomol. 85: 371-374.

Lee, R. E., J. M. Strong-Gunderson, M. R. Lee, K. S. Grove & T. J. Riga. 1991. Isolation of ice nucleating active bacteria from insects. J. Exp. Zool. 257: 124-127.

Lindow, S. E. 1983. The role of bacterial ice nucleation in frost injury to plants. Ann. Rev. Phytopathol. 21: 363-384.

Lindow, S. E. 1995. Control of epiphytic ice nucleation-active bacteria for managment of plant frost injury, *In* R. E. Lee, G. J. Warren & L. V. Gusta, eds. *Biological Ice Nucleation and Its Applications*. pp 239-256. American Phytopathol. Soc., St. Paul, Minnesota.

Lindow, S. E., D. C. Arny, W. R. Barchet & C. D. Upper. 1978. The role of bacterial ice nuclei in frost injury to sensitive plants, *In* P. Li, ed. *Plant Cold Hardiness and Freezing Stress*. pp 249-263. Academic Press, New York.

Mail, G. A. & R. W. Salt. 1933. Temperature as a possible limiting factor in the northern spread of the Colorado potato beetle. J. Econ. Entomol. 26: 1068-1075.

Milner, M., K.-J. S. Kung, J. A. Wyman, J. Feldman & E. Nordheim. 1992. Enhancing overwintering mortality of Colorado potato beetle (Coleoptera: Chrysomelidae) by manipulating the temperature of its habitat. J. Econ. Entomol. 85: 1701-1708.

Mugnano, J.A., R.E. Lee & R.T. Taylor. 1996. Fat body cells and calcium phosphate spherules induce ice nucleation in the freeze-tolerant larvae of the gall fly, *Eurosta solidaginis* (Diptera, Tephritidae). J. Expt. Biol. 199: 465-471.

Pouleur, S., C. Richard, J.-G Martin & H. Antonn. 1992. Ice nucleation activity in *Fusarium acuminatum* and *Fusarium avenaceum*. Appl. Environ. Microbiol. 58:2960-2964.

Ring, R. A. & H. V. Danks. 1994. Desiccation and cryoprotection: overlapping adaptations. Cryo-Letters. 15: 181-190.

Sømme, L. 1982. Supercooling and winter survival in terrestrial arthropods. Comp. Biochem. Physiol. 73A: 519-543.

Sømme, L. 1989. Adaptations of terrestrial arthropods to the alpine environment. Biol. Rev. 64: 367-407.

Steigerwald, K. A., M. R. Lee, R. E. Lee & J. C. Marshall. 1995. Effect of biological ice nucleators on insect supercooling capacity varies with anatomic site of application. J. Insect Physiol. 41: 603-608.

Storey, J. M. & K. B. Storey. 1988. Freeze tolerance in animals. Physiol. Rev. 68: 27-84.

Strong-Gunderson, J. M., R. E. Lee & M. R. Lee. 1992. Topical application of ice-nucleating-active bacteria decreases insect cold tolerance. Appl. Env. Microbiol. 58: 2711-2716.

Strong-Gunderson, J. M., R. E. Lee, M. R. Lee & T. J. Riga. 1990. Ingestion of ice-nucleating active bacteria increases the supercooling point of the lady beetle *Hippodamia convergens*. J. Insect Physiol. 36: 153-158.

Tauber, M. J., C. A. Tauber & S. Masaki. 1986. *Seasonal Adaptations of Insects* New York: Oxford University Press. 411 pp.

Tauber, M. J., C. A. Tauber & J. P. Nyrop. 1994. Soil moisture and postdormancy of Colorado potato beetles (Coleoptera: Chrysomelidae): Descriptive model and field emergence patterns. Environ. Entomol. 23: 1485-1496.

Tsumuki, H. 1992. An ice-nucleating active fungus isolated from the gut of the rice stem borer, *Chilo suppressalis* Walker (Lepidoptera: Pyralidae). J. Insect Physiol. 38: 119-125.

Tsumuki, H. & H. Konno. 1991. Tissue distribution of the ice-nucleating agents in larvae of the rice stem borer, *Chilo suppressalis* Walker (Lepidoptera: Pyralidae). Cryobiol. 28: 376-381.

Upper, C. D. & G. Vali. 1995. The discovery of bacterial ice nucleation and its role in the injury of plants by frost, *In* R. E. Lee, G. J. Warren & L. V. Gusta, eds. *Biological Ice Nucleation and Its Applications.* pp 29-39. American Phytopathol. Soc., St. Paul, Minnesota.

Ushatinskaya, R. S. 1978. Seasonal migration of adult *Leptinotarsa decemlineata* (Insecta, Coleoptera) in different types of soil and physiological variations of individuals in hibernating populations. Pedobiologia, Bd. 18: 120-126.

Vali, G. 1971. Quantitative evaluation of experimental results on the heterogeneous freezing nucleation of supercooled liquids. J. Atmos. Sci. 28: 402-409.

Vali, G. 1995. Principles of ice nucleation, *In* R. E. Lee, G. J. Warren & L. V. Gusta, eds. *Biological Ice Nucleation and Its Applications.* pp 1-28. American Phytopathol. Soc., St. Paul, Minnesota.

Warren, G. J. 1995. Identification and analysis of *ina* genes and proteins, *In* R. E. Lee, G. J. Warren & L. V. Gusta, eds. *Biological Ice Nucleation and Its Applications.* pp. 85-99. American Phytopathol. Soc., St. Paul, Minnesota.

Wasylyk, J. M., A. Tice & J. G. Baust. 1988. Partial glass formation: a novel mechanism of insect cryoprotection. Cryobiol. 25: 251-258.

Weber, D. C. & D. N. Ferro. 1993. Distribution of overwintering Colorado potato beetle in and near Massachusetts potato fields. Entomol. exp. appl. 66: 191-196.

Wolber, P. K. 1993. Bacterial ice nucleation. Adv. Microb. Physiol. 34: 203-237.

Wolber, P. K., R. L. Green, W. T. Tucker, N. M. Watanabe, C. A. Vance, R. A. Fallon, C. Lindhardt & A. J. Smith. 1995. Identification and analysis of ina genes and proteins, in R. E. Lee, G. J. Warren & L. V. Gusta, eds. *Biological Ice Nucleation and Its Applications.* pp 283-298. American Phytopathol. Soc., St. Paul, Minnesota: .

Zachariassen, K. E. 1985. Physiology of cold tolerance in insects. Physiol. Rev. 65: 799-832.

Zachariassen, K. E. 1992. Ice nucleating agents in cold-hardy insects, *In* G. N. Somero, C. B. Osmond & C. L. Bolis, eds. *Water and Life.* pp 261-281. Berlin: Springer-Verlag.

Zachariassen, K. E. & H. T. Hammel. 1976. Nucleating agents in the haemolymph of insects tolerant to freezing. Nature 262: 285-287.

Zachariassen, K. E. & R. Lundheim. 1992. The endocrine control of insect cold hardiness. Zool. Jb. Physiol. 96: 183-196.

5

Temperature Synergism in Integrated Pest Management

David J. Horn

The influence of temperature on biochemical reactions and resultant arthropod activity may either enhance or limit the effectiveness of integrated pest management (IPM). Feeding, dispersal, and reproductive rates of pest arthropods and their natural enemies generally increase with increasing temperatures through the range 10° to 40°C although variation in response to temperatures within this range may lead to a wide array of outcomes. Temperature also influences the expression of genes within plants and in microbial pathogens, and some genes that code for plant allelochemicals or microbial toxins may be active at only high or low temperatures. Temperature influences both physical and chemical properties of insecticides such as volatilization and breakdown.

These temperature-mediated synergisms add complexities that are not always appreciated by those devising or using various IPM strategies. Effects of extreme temperatures can lead to failure of an IPM strategy that may be quite effective in a narrow temperature range. A pest management technique effective over a limited temperature range results in outlying temperatures to which a pest population may "escape" into a warmer or cooler location or time where management is ineffective. A prudent pest manager must appreciate that the complex interaction of temperature with other ecosystem components can enhance or retard the effectiveness of IPM.

Central to the implementation of any IPM system is the concept of the economic injury level: the population density of a pest at which control measures must be undertaken to avoid economic losses (Higley & Pedigo 1996). Growth of an arthropod population is a function of temperature, and at warmer temperatures, pest populations grow faster and thus reach economic injury levels sooner. Most predictive models of economic injury levels take this into account although accurate temperature measurement and forecasting are essential for successful application of these models to the real world (Horn 1988, Higley & Pedigo 1996). At best, climatological data taken according to standard observational procedures are an approximation of the conditions actually experienced by insects, particularly in the soil.

At extreme temperatures, arthropods and their host plants may be differentially stressed. Outbreaks of spider mites (Tetranychidae) are a classic example. In the midwestern United States, during hot dry midsummer weather, apples or soybeans may suffer heavier than usual damage because spider mites (*Panonychus ulmi* on apples and *Tetranychus urticae* on soybeans) reproduce at their maximum rates and the host plants may be heat and water stressed as well. Producers of greenhouse crops are quite aware of the increased impact of spider mites at high temperatures when generation times are reduced, feeding intensity is increased, and plants are showing water stress. The locomotory speed and dispersal ability of the greenhouse whitefly, *Trialeurodes vaporariorum*, is temperature dependent (Noldus & van Lenteren 1989), and therefore whiteflies spread through greenhouses much faster in warmer weather. The interrelationship between temperature and plant growth is such that if plants develop faster at lower temperatures than do insects, the economic injury level may not be reached (Lerin & Koubaiti 1995). The direct influence of temperature on pest population growth may be altered by changes in light intensity or photoperiod (Wyatt & Brown 1977). Biological control agents, especially arthropod natural enemies, often exhibit temperature optima different from those of their prey, and may become ineffective at higher or lower temperatures. Extreme temperatures may impact the expression of certain genes in plants, or may alter enzyme systems, thus altering effectiveness of host plant resistance. Finally, prolonged exposure to a prevailing temperature may alter an insect's behavior; Myers & Sabath (1981) found that the threshold temperature for emergence of adult cinnabar moth, *Tyria jacobeae*, was changed after 17 generations of exposure to lower temperatures in the laboratory. This chapter explores some examples and consequences of temperature interactions with pest management strategies such as chemical and biological control, host plant resistance, and the use of pheromones.

Insecticides

Chemical insecticides are often the only effective remedy for quickly and inexpensively reducing pest populations to below economic injury levels. Stability, vaporization, penetration, activity, and degradation of insecticides are all partly dependent on physical and biochemical processes that proceed at characteristic rates at different temperatures. Generally these reactions proceed more rapidly at higher temperatures although not always do they show a linear relationship with temperature. Some insecticides are more efficient in both toxicity and knockdown at lower temperature; pyrethroid activity against the house fly is greater at 18°C than at 32°C (Table 5.1). Such a relationship is termed Negative Temperature Activity Coefficient (NTAC) and is a general characteristic of pyrethroid insecticides (Wadleigh et al. 1991). Positive Temperature Activity Coefficient (PTAC) denotes the opposite effect, often characteristic of organophosphates (Table 5.2). For example, chlorpyrifos vaporization and penetration are greater at 24°C than at 12°C in the laboratory. If this occurs in the field, there will be shorter residual activity along with increased penetration into insects at higher temperatures (Hill et al. 1996). The realized efficacy of an insecticide is a function of the interrelationship between vaporization and cuticular penetration. Residual activity of an insecticide is usually correlated with temperature. Generally, an insecticide will remain active for longer when temperatures are lower, due mostly to reduced vaporization. Also, microbial degradation of insecticide is

TABLE 5.1 Toxicity (LD_{50} at 24 h) and mortality/knockdown ratio (MKR) of four pyrethroids at three temperatures in the house fly, *Musca domestica* (Scott & Georghiou 1984).

Insecticide	18 °C		25 °C		32 °C	
	LD_{50}	MKR	LD_{50}	MKR	LD_{50}	MKR
allethrin	355.7	10.7	455.1	5.33	414.2	3.36
cypermethrin	3.27	9.70	7.37	9.70	31.7	18.7
flucythrinate	29.3	13.8	62.5	11.4	264.0	28.3
permethrin	28.3	2.91	53.5	2.67	101.3	2.17

TABLE 5.2 Toxicity (LD_{50} at 72 h) of three insecticides to three lepidopterans at three temperatures (Sparks et al. 1982).

Insect	Insecticide	Temperature(°C)	LD_{50}
Trichoplusia ni	permethrin	37.8	0.540
		26.7	0.244
		15.6	0.072
	fenvalerate	37.8	0.228
		26.7	0.168
		15.6	0.112
	deltamethrin	37.8	0.016
		26.7	0.012
		15.6	0.005
Spodoptera frugiperda	permethrin	37.8	1.084
		26.7	0.733
		15.6	0.208
	fenvalerate	37.8	6.864
		26.7	10.316
		15.6	8.248
	deltamethrin	37.8	0.180
		26.7	0.180
		15.6	0.340
Heliothis virescens	permethrin	37.8	1.944
		26.7	1.440
		15.6	0.216
	fenvalerate	37.8	0.224
		26.7	0.396
		15.6	0.512
	deltamethrin	37.8	0.016
		26.7	0.044
		15.6	0.088

inhibited by low temperatures, resulting in longer residual efficacy (Felsot 1989).

An insect may be more susceptible to insecticide due to increased impact of heat. Moss & Jang (1991) showed that rotenone and dinitrophenol increased heat sensitivity in the Mediterranean fruit fly, *Ceratitis capitata*. Increased concentrations of CO_2 at high temperatures (45 °C) resulted in more rapid kill of codling moth, *Cydia pomonella*, larvae on apples in storage when exposed to CO_2 (Soderstrom et al. 1996). Temperature may synergize activity of an insecticide, or an insecticide may enhance the impact of high temperature. Ebeling (1990) demonstrated this by exposing confused flour beetles, *Tribolium confusum*, and German cockroaches, *Blattella germanica*, to boric acid powder at 46 °C; the mortality due to heat and boric acid together was greater than that due to either factor alone. By cooling grain to 15 °C, Longstaff (1988) achieved 80% reduction in amount of pyrethroid applied to achieve control of the rice weevil, *Sitophilus oryzae*, although this was not the case when organophosphate insecticides were used. Johnston & Corbett (1986) showed that increased activity of fenitrothion from 17° to 22 °C was due to enzymatic activity increasing the concentration of the more toxic metabolite fenitrooxon. The impact of inorganic silica on confused flour beetles is enhanced by high temperature (Ebeling 1995).

Temperature may interact with the expression of pesticide resistance, and temperature effects should be considered in resistance management. Patil et al. (1996) noted that pre-stressing larvae of *Anopheles stephensi* and *Aedes aegypti* with high temperatures (43 °C) resulted in enhanced cross-resistance to propoxur, and conversely, exposure to propoxur enhanced the mosquitoes' heat tolerance. Tolerance of citrus thrips, *Scirtothrips citri*, to acephate and fluvalinate decreased as temperatures increased from 16° to 32 °C (Table 5.3). A field population of *Drosophila melanogaster* selected for cyclodiene resistance displayed a temperature-related paralysis at high temperature (38 °C) in the laboratory, and such an effect could reduce fitness at high temperatures (ffrench-Constant et al. 1993). Brown (1987) noted that a pyrethroid-tolerant strain of *Heliothis virescens* displayed resistance to fenvalerate, fluocythrinate, and permethrin at 26 °C but not at 16 °C, indicating that resistance in the field could go unnoticed at low temperatures. Foster et al. (1996) reported apparent loss of insecticide resistance in the peach aphid, *Myzus persicae*, after exposure to low temperature and theorized that selection for resistance could be negated by cold winter weather. The biochemical mechanisms responsible for these effects have not been thoroughly investigated.

TABLE 5.3 Effect of temperature on tolerance of citrus thrips to acephate and fluvalinate (Zareh & Morse 1989).

Insecticide	Mean temp.(°C)	LD_{50}	LD_{90}
acephate	16.1	8.879	72.204
	22.0	7.636	39.090
	28.0	3.922	40.849
	32.0	2.187	14.239
fluvalinate	16.1	0.433	2.645
	22.0	0.271	1.430
	28.0	0.186	1.830
	32.0	0.030	0.424

Biological Control

In its traditional ("classical") sense, biological control is the use of arthropod natural enemies (parasitoids and predators) for control of noxious insect pests. The aim of biological control is to establish a lower equilibrium population of a pest, thus reducing, or eliminating the need for additional manipulations such as chemical insecticides.

Developmental times of prey and natural enemies are correlated with temperature and especially at higher temperatures, prey may mature faster and their populations therefore may increase faster than the predators' ability to respond numerically. The relationship between the developmental rate of a predator or parasitoid and that of its prey or host may be critical to the success of biological control (Bernal & Gonzalez 1993, 1996). For example, when the alfalfa weevil initially invaded coastal California it was partially controlled by the larval ichneumonid parasitoid *Bathyplectes curculionis*. Once the weevil became established in the hotter, drier San Joaquin Valley, the parasitoid, although it was abundant, could no longer keep pace with the host population (Michelbacher 1940). The spider mite predator *Phytoseiulus persimilis* is a very effective biological control agent against the two-spotted spider mite, *Tetranychus urticae*, in greenhouses at temperatures up to 30°C. Beyond that temperature, *P. persimilis* seeks cooler portions of plants. In North American greenhouses, which are generally hotter than those of western Europe, *T.*

urticae "escapes" to the tops of the plants where not only are there no predators present, but temperatures are suitable for maximum reproduction. The result is continuous production of hungry phytophagous spider mites from the predator-free space above the lower leaves (Helle & Sabelis 1985). The predators remain abundant but ineffective in control. Miller (1996) found that the threshold temperature for development of the braconid parasitoid *Meteorus communis* was greater than that of its host the green cloverworm, *Plathypena scabra*, in Oregon, in contrast to the eastern USA where host and parasitoid expressed about the same developmental threshold. As a result, parasitism by *M. communis* is a minimal factor in biological control of the green cloverworm in Oregon. Messenger & van den Bosch (1971) suggested that practitioners of biological control should attempt to match agents from areas of climate similar to that occupied by the target species. Alternatively, Roush (1989) recommended that biological control agents should be selected for tolerance to a maximum temperature range in order to avoid the difficulty of incompatible climates. The initial release of *Tetrastichus incertus* for alfalfa weevil control in the eastern USA was of the progeny from a single host from France, and the parasitoid did not fare well initially in upstate New York due in part to the rigors of cold winter temperatures (Horn 1971). A Scandinavian strain of *T. incertus* was later introduced with greater success. The reproductive rate of the aphidiid parasitoid *Lysiphlebus testaceipes* may be too low for it to overtake *Aphis gossypii* at most temperatures (van Steenis 1994). Exposure to low winter temperatures apparently inhibits the effectiveness of the parasitoid *Aphelinus albipodus* against its host, the Russian wheat aphid, *Diuraphis noxia*, at temperatures less than 10°C. Low temperatures may induce diapause in a biological control agent but not in its pest prey (van Houten et al. 1988). Temperatures less than 12°C induced diapause in the predatory mite *Amblyseius cucumeris*, rendering it ineffective in management of thrips on outdoor crops in cool weather (Rodriguez-Reina et al. 1994).

Microbial pathogens offer an environmentally benign alternative to chemical insecticides. The infection rate and subsequent reproduction of microbial pathogens are impacted by temperature with generally increased activity at higher temperatures. Infection of blackmargined aphid, *Monellia caryella*, by three entomophthoran fungi needs temperatures greater than 8°C (Ekbom & Pickering 1990). Ribeiro & Pavan (1994) noted that increase in infectivity of nuclear polyhedrosis viruses (NPV) can be enhanced by laboratory selection with high temperatures; they found a tenfold increase in yield of NPV polyhedra in *Diatraea saccharalis* larvae after such a selection protocol (Table 5.4). Johnson et al. (1982) developed a successful temperature-dependent model for infestation by NPV in the velvetbean caterpillar, *Anticarsia gemmatalis,* and this is useful in determining whether application will be successful in the field.

With the advent of bioengineered plants containing insect toxins, researchers who conduct bioassays have to consider the effects of temperature on gene expression and quantity of toxin produced. Robertson et al. (1996) reported that a bioengineered *Bacillus thuringiensis* preparation was less effective than a commercial preparation (Dipel) at 25°C although this difference disappeared at 28°C.

Host Plant Resistance

Planting of insect-resistant cultivars of crop plants is a time-honored IPM technique particularly in field crops such as maize, wheat, and alfalfa, to which insecticide application is often not economical. More recently, host plant resistance is increasingly considered a pest management option for producers

TABLE 5.4 Effect of temperature on mortality of third instar *Diatraea saccharalis* fed unselected and temperature-selected strains of two nuclear polyhedrosis viruses (Ribeiro & Pavan 1994).

	Percentage mortality			
Temperature(°C)	Ag *Virus*		Tn *Virus*	
	unselected	selected	unselected	selected
17	7	47	35	75
22	12	90	45	85
24	10	92	50	80
26	12	95	45	85
28	27	97	62	100
30	50	97	65	100
32	50	97	80	100
35	70	100	95	100
37	100	100	100	100
39	100	100	100	100

of high value crops like fruits, vegetables, and ornamentals. The effectiveness of host plant resistance as a pest management tactic is often a function of temperature, and plants may resist insect infestations over a temperature range more limited than the crop experiences during the growing season (Tingey & Singh 1980). This may vary according to both the cultivar of plant and the insect pest biotype. Cartwright et al. (1946) demonstrated this initially in wheat varieties resistant to different biotypes of the Hessian fly, *Mayetiola destructor*. Low or high temperature stress within plants can reduce output of allelochemicals resulting in apparent susceptibility. Salim & Saxena (1991) reported this for planthoppers on rice. Stamp (1993) noted that the interaction between temperature and plant quality may be very complex.

As noted earlier with insecticides, some plant cultivars may express increased resistance with rising temperatures. Schalk et al. (1969) reported that aphid-resistant alfalfa lost resistance to spotted alfalfa aphid, *Therioaphis maculata*, and pea aphid, *Acyrthosiphum pisum*, after exposure to temperatures of 10 to 15°C. Resistance of wheat to greenbug, *Schizaphis graminum*, Biotype I was significantly reduced at low temperature in cultivar Cargill 607E, and it is presumed that low temperature negatively impacts expression of antibiosis (Harvey et al. 1994). Stamp & Yang (1996) found that tomatine reduced the developmental rate of *Helicoverpa zea* more markedly at higher temperatures. Wood & Starks (1971) noted increased resistance to greenbug in sorghum as temperature increased. Expression of tolerance and antixenosis of sorghum to greenbug biotypes C and E was greater at 30° than at 26° (Thindwa & Teetes 1994).

In others cases plants may lose resistance as temperature increases. Alfalfa varieties resistant to spotted alfalfa aphid are more effective at lower temperatures, apparently due in part to the aphids' propensity for dispersing onto susceptible cultivars at temperatures around 10°C (Kindler & Staples 1970). The actual expression of resistance by alfalfa cultivars to spotted alfalfa and pea aphids appears to decrease with lower temperatures (Isaak et al. 1963). Tyler & Hatchett (1983) reported loss of Hessian fly resistance in some wheat cultivars at temperatures exceeding 18°C. Sosa & Foster (1976) suggested that researchers should screen wheat varieties at temperatures exceeding 27°C in order to account for variability in resistance among different cultivars. Fluctuating temperatures may affect expression of resistance more extensively than do constant temperatures (Hagstrum & Milliken 1991).

Loss of the effectiveness of resistance at higher temperatures may result simply from the differential impact of temperature on plant vs. insect growth.

As noted earlier, most insect pests grow faster and reproduce earlier at higher temperatures, and may simply overwhelm the plants' capacity to produce the chemical components that confer resistance. The differential growth rates between plants and phytophagous arthropods may thus be interpreted as an inhibition of resistance (Stiefel et al. 1992). Failure of host plant resistance may be due to more subtle mechanisms. Walters et al. (1991) reported loss of resistance to aphids and whiteflies in geraniums at high temperatures was due to increased production of short chain anacardic acid within leaf trichomes. The resulting fluid was less viscous and did not entrap the insects nearly as well as the more viscous fluid produced at cooler temperatures.

Pheromones

Pheromones, particularly sex pheromones, are finding increasing use in IPM, primarily to monitor insect activity for more efficient timing of pesticide applications. In a few instances, synthetic pheromones are applied to plants in order to reduce the frequency of successful mating. This so-called "confusion technique" has been notably successful against the pink bollworm, *Pectinophora gossypiella*.

Low or high temperatures suppress activity in many insect species, and this can limit the interpretation of pheromone trap catches in monitoring insect populations. Schouest & Miller (1994) found that night temperatures below $20\,^{\circ}C$ suppressed moth flight, and they recommended that weather data be recorded wherever automated pheromone traps were in use to assure accurate interpretation of moth counts from traps. High temperature increases volatilization of pheromones, resulting in increased effectiveness of pheromone-baited traps and greater success of the confusion technique when pheromone is applied to a crop plant (Flint et al. 1990). Jonsson & Anderbrant (1993) reported large increases of the European pine sawfly, *Neodiprion sertifer*, in pheromone traps at high temperatures. Conversely, temperatures above $35\,^{\circ}C$ result in a decline of pheromone titer in female gypsy moths, *Lymantria dispar* (Giebultowicz et al. 1992). Lextrait et al. (1995) noted that pheromone emission by fully-winged and flight worthy females of *Callosobruchus maculatus* was enhanced at high temperatures but that the pheromone output of flightless females was unaffected. Marks (1977) reported that there was no simple relationship between temperature and responsiveness to pheromone traps in the red bollworm although male flight activity peaked earlier in the evening if temperature was lower. Pheromone trap catches thus were unreliable as an accurate forecasting tool. Delisle & McNeil (1987) found that the onset of calling in the armyworm, *Pseudaletia unipuncta*, was advanced when the moths were cooled from 25 successively to 20, 15, and 10°C, and duration of calling also increased, shifting the "calling gate" earlier in response to low night temperatures. Cardé et al. (1996) found a complex relationship between temperature and calling activity in the gypsy moth. Warm daytime temperatures increased the numbers of males found in pheromone traps, and low temperatures suppressed the activity of the first diel peak of male activity. A second peak in male responsiveness to pheromone traps was unaffected except that it occurred later in the evening at lower temperatures. The time of day when pheromone traps are checked for male moths thus may be a factor in interpretation of moth numbers. McNally & van Steenwyck (1986) found the relationship between temperature at sunset and trap catches to be sufficiently complex that there was no predictable relationship between pheromone catches and level of larval infestation level in walnuts in most years. In this case, trap catches were a marginally reliable predictor of infestation.

Future Directions

Integrated Pest Management systems need to make allowance for the impact of temperature, especially when the impact of temperature is more subtle and complex than merely a linear relationship with biochemical activity. Insecticide screening is normally done over the entire range of temperatures under which the insecticide is likely to be used, and this advice could apply equally to those who screen bioengineered plants. The numerous examples of variability in the expression of host plant resistance are testimony to the likelihood that bioengineered crops are likely to respond to temperature in a complex fashion. Likewise, biological or microbial control agents should be selected with regard to a broad range of temperature conditions so that they can be effective over the maximum of the target's geographical distribution. We need to be aware that a pest arthropod may itself alter its temperature optima over time in response to selection. The biochemical basis of insecticide resistance should be investigated in much greater detail. It is especially important to learn the specific enzyme pathways responsible for detoxification, the temperature optima and limits for the activity of these enzyme systems, and their relationship to the ambient temperatures experienced by the pest population during a growing season. In integrating various control tactics into a total IPM system overall, designers of IPM systems need to consider the entire range of possible temperatures and how temperature is likely to interact with each system component as well as with the total system. More thorough attention to temperature synergisms should lead to more effective IPM over a wider range of environmental conditions.

References

Bernal, J. S. & D. Gonzalez. 1993. Temperature requirements of four parasites of the Russian wheat aphid *Diuraphis noxia*. Entomol. Exp. Appl. 69: 173-182.

Bernal, J. S. & D. Gonzalez. 1996. Thermal requirements of *Aphelinus albipodus* (Hayat and Fatima) (Hym., Aphelinidae) on *Diuraphis noxia* (Mordmilko) (Hom., Aphididae) hosts. J. Appl. Entomol. 120: 631-638.

Brown, M. A. 1987. Temperature-dependent pyrethroid resistance in a pyrethroid-selected colony of *Heliothis virescens* (F.) (Lepidoptera: Noctuidae). J. Econ. Entomol. 80: 330-332.

Cardé, R. T., R. E. Charlton, W. E. Wallner & Y. N. Baranchikov. 1996. Pheromone-mediated diel activity rhythms of male Asian gypsy moth (Lepidoptera: Lymantriidae) in relation to female eclosion and temperature. Ann. Entomol. Soc. Amer. 89: 745-753.

Cartwright, W. B., R. M. Caldwell & L. E. Compton. 1946. Relationship of temperature to the expression of resistance in wheats to Hessian fly. J. Amer. Soc. Agron. 38: 259-263

Delisle, J. & J. N. McNeil. 1987. Calling behaviour and pheromone titre of the true armyworm, *Pseudaletia unipuncta* (Haw.) (Lepidoptera: Noctuidae) under different temperature and photoperiodic conditions. J. Insect Physiol. 33: 315-324.

Ebeling, W. 1990. Heat and boric acid: a striking example of synergism. Pest Contr. Technol. 18: 44-46.

Ebeling, W. 1995. Inorganic insecticides and dusts. *In* M. K. Rust, J. M. Owens & D. A. Reierson, eds. *Understanding and Controlling the German Cockroach*, pp 193-230. Oxford Univ. Press, New York.

Ekbom, B. S. & J. Pickering. 1990. Pathogenic fungal dynamics in a fall population of the blackmargined aphid (*Monellia caryella*). Entomol. Exp. Appl. 57: 29-37.

Felsot, A. S. 1989. Enhanced biodegradation of insecticides in soil: implications for agroecosystems. Annu. Rev. Entomol. 34: 453-476.

ffrench-Constant, R. H., J. C. Steichen & P. J. Ode. 1993. Cyclodiene insecticide resistance in *Drosophila melanogaster* (Meigen) is associated with a temperature-sensitive phenotype. Pest. Biochem. Physiol. 46: 73-77.

Flint, H. M., A. Yamamoto, N. J. Parks & K. Nyomura. 1990. Aerial concentrations of gossyplure, the sex pheromone of the pink bollworm (Lepidoptera:

Gelechiidae), in cotton fields treated with long-lasting dispensers. Envir. Entomol. 19: 1845-1851.

Foster, S. P., R. Harrington, A. L. Devonshire, I. Denholm, G. J. Devine & M. G. Kenward. 1996. Comparative survival of insecticide-susceptible and resistant peach-potato aphids, *Myzus persicae* (Sulzer) (Hemiptera: Aphididae), in low temperature field trials. Bull. Entomol. Res. 86: 17-27.

Giebultowicz, J. M., R. E. Webb, A. K. Raina & R. L. Ridgway. 1992. Effects of temperature and age in daily changes in pheromone titre in laboratory-reared and wild gypsy moth (Lepidoptera: Lymantriidae). Envir. Entomol. 21: 822-826.

Hagstrum, D. W. & G. A. Milliken. 1991. Modeling differences in insect developmental times between constant and fluctuating temperatures. Ann. Entomol. Soc. Amer. 84: 369-379.

Harvey, T. L., G. E. Wilde, K. D. Kofoid & P. J. Bramel-Cox. 1994. Temperature effects on resistance to greenbug (Homoptera: Aphididae) biotype I in sorghum. J. Econ. Entomol. 87: 500-503.

Helle, W. & M. Sabelis. 1985. *Spider Mites: Their Biology, Natural Enemies and Control*. Elsevier, Amsterdam.

Hill, B. D., R. A. Butts & G. B. Schaalje. 1996. Factors affecting chlorpyrifos activity against Russian wheat aphid (Homoptera: Aphididae) in wheat. J. Econ. Entomol. 89: 1004-1009.

Higley, L. G. & L. P. Pedigo eds. 1996. *Economic Thresholds for Integrated Pest Management*. Univ. Nebraska Press, Lincoln NE.

Horn, D. J. 1971. The relationship between a parasite, *Tetrastichus incertus* (Hymenoptera: Eulophidae) and its host, the alfalfa weevil, *Hypera postica* (Coleoptera: Curculionidae) in New York. Canad. Entomol. 103: 83-94.

Horn, D. J. 1988. *Ecological Approach to Pest Management*. Guilford, New York. 285 pp.

Isaak, A., E. L. Sorensen & E. E. Ortman. 1963. Influence of temperature and humidity on resistance in alfalfa to the spotted alfalfa aphid and pea aphid. J. Econ. Entomol. 56: 53-57.

Johnson, D. W., D. B. Boucias, C. S. Barfield & G. E. Allen. 1982. A temperature-dependent model for a nuclearpolyhedrosis virus of the velvetbean caterpillar, *Anticarsia gemmatalis* (Lepidoptera: Noctuidae). J. Invert. Pathol. 40: 292-298.

Johnston, J. J. & M. D. Corbett. 1986. The effects of salinity and temperature on the *in vitro* metabolism of the organophosphorus insecticide fenitrothion by the blue crab, *Callinectes sapidus*. Pest. Biochem. Physiol. 26: 193-201.

Jonsson, P. & O. Anderbrant. 1993. Weather factors influencing catch of *Neodiprion sertifer* (Hymenoptera: Diprionidae) in pheromone traps. Envir. Entomol. 22: 445-452.

Kindler, S. D. & R. Staples. 1970. The influence of fluctuating and constant temperatures, photoperiod, and soil moisture on the resistance of alfalfa to the spotted alfalfa aphid. J. Econ. Entomol. 63: 1198-1201.

Lerin, J. C. & K. Koubaiti. 1995. Effect of temperature and plant size on the infestation dynamics of oilseed rape plants by *Baris coeruelescens* Scop. (Col., Curculionidae) in field conditions. J. Appl. Entomol. 119: 149-156.

Lextrait, P., J. C. Biemont & J. Pouzat. 1995. Pheromone release by the two forms of *Callosobruchus maculatus* females: efects of age, temperature and host plant. Physiol. Entomol. 20: 309-317.

Longstaff, B. C. 1988. A modelling study of the effects of temperature manipulation upon the control of *Sitophilus oryzae* (Coleoptera: Curculionidae) by insecticide. J. Appl. Ecol. 25: 163-175.

Marks, R. J. 1977. The influence of climatic factors on catches of the red bollworm *Diparopsis castanea* Hampson (Lepidoptera: Noctuidae) in sex pheromone traps. Bull. Entomol. Res. 67: 243-248.

McNally, P. S. & R. Van Stenwyck. 1986. Relationship between pheromone-trap catches and sunset temperatures during the spring flight and codling moth (Lepidoptera: Olethreutidae) infestations in walnuts. J. Econ. Entomol. 79: 444-446.

Messenger, P. S. & R. van den Bosch. 1971. The adaptability of introduced biological control agents. *In* C. B. Huffaker, ed. *Biological Control*, pp 68-92. Plenum, New York.

Michelbacher, A. E. 1940. Effect of *Bathyplectes curculionis* on the alfalfa weevil population in lowland middle California. Hilgardia 13: 81-99.

Miller, J. C. 1996. Temperature-dependent development of *Meteorus communis* (Hymenoptera: Braconidae), a parasitoid of the variegated cutworm (Lepidoptera: Noctuidae). J. Econ. Entomol. 89: 877-880.

Miller, J. C. & W. Gerth. 1994. Temperature-dependent development of *Aphidius matricariae* as a parasitoid of the Russian wheat aphid. Environ. Entomol. 23: 1304-1307.

Moss, J. I. & E. B. Jang. 1991. Effects of age and metabolic stress on heat tolerance of Mediterranean fruit fly (Diptera: Tephritidae) eggs. J. Econ. Entomol. 84: 537-541.

Myers, J. H. & M. D. Sabath. 1981. Genetic and phenotypic variability, genetic variation, and the success of establishment of insect introductions for the biological control of weeds. *In* E. S. DelFosse, ed. *Proceedings of the Fifth International Symposium for Biological Control of Weeds*, pp. 91-102. CSIRO, Canberra, Australia.

Noldus, L. P. J. J. & J. C. Van Lenteren. 1989. Host aggregation and parasitoid behaviours: biological control in a closed system. *In* M. Mackauer, L. E. Ehler & J. Roland, eds. *Critical Issues in Biological Control*, pp 229-262. Intercept, Ltd. Andover, U.K.

Patil, N. S., K. S. Lole & D. N. Deobagkar. 1996. Adaptive larval thermotolerance and induced cross-tolerance to propoxur insecticide in mosquitoes *Anopheles stephensi* and *Aedes aegypti*. Med. Vet. Entomol. 10: 277-282.

Ribeiro, H. C. T. & O. H. O. Pavan. 1994. Effect of temperature on development of baculoviruses. J. Appl. Entomol. 118: 316-320.

Robertson, J. L., H. K. Preisler, S. L. Ng, L. A. Hickle, A. Berdeja & W. D. Gelernter. 1996. Comparative effect of temperature and time on activity of Dipel 2X and MVP preparation of *Bacillus thuringiensis* subsp. *kurstaki* on diamondback moth (Lepidoptera: Plutellidae). J. Econ. Entomol. 89: 1084-1087.

Rodriguez-Reina, J. M., F. Ferragut, A. Carnero & M. A. Pena. 1994. Diapause in the predacious mites *Amblyseius cucumeris* (Oud.) and *A. barkeri* (Hug.): consequences of use in integrated control programmes. J. Appl. Entomol. 118: 44-50.

Roush, R. T. 1989. Genetic variation in natural enemies: critical issues for colonization in biological control, *In* M. Mackauer, L.E. Ehler & J. Roland, eds. *Critical Issues in Biological Control*, pp 263-288 Intercept, Ltd. Andover, U.K.

Salim, M. & R. C. Saxena. 1991. Temperature stress and varietal resistance in rice: effects on whitebacked planthopper. Crop Sci. 31: 1620-1625.

Schalk, J. M., S. D. Kindler & G. R. Manglitz. 1969. Temperature and the preference of the spotted alfalfa aphid for resistant and susceptible alfalfa plants. J. Econ. Entomol. 62: 1000-1003.

Schouest, L. P., Jr. & T. A. Miller. 1994. Automated pheromone traps show male pink bollworm (Lepidoptera: Gelechiidae) mating response is dependent on weather conditions. J. Econ. Entomol. 87: 965-974.

Scott, J. G. & G. Georghiou. 1984. Influence of temperature on knockdown toxicity and resistance to pyrethroids in the house fly, *Musca domestica*. Pest. Biochem. Physiol. 21: 53-62.

Soderstrom, E. L., D. G. Brandl & B. E. Mackey. 1996. High temperature alone and combined with controlled atmospheres for control of diapausing codling moth (Lepidoptera: Tortricidae) in walnuts. J. Econ. Entomol. 89: 144-147.

Sosa, O. Jr. & J. E. Foster. 1976. Temperature and the expression of resistance in wheat to the Hessian fly. Envir. Entomol. 5: 333-336.

Sparks, T. C., M. H. Shour & E. G. Wellemeyer. 1982. Temperature-toxicity relationships of pyrethroids on three lepidopterans. J. Econ. Entomol. 75: 643-646.

Stamp, N. E. 1993. A temperate region view of the interaction of temperature, food quality and predators on caterpillar foraging, *In* N. E. Stamp & T. M. Casey, eds. *Caterpillars: Ecological and Evolutionary Constraints on Foraging*, pp 478-508. Chapman & Hall, New York.

Stamp, N. E. & Y. Yang. 1996. Response of insect herbivores to multiple allelochemicals under different thermal regimes. Ecology 77: 1088-1102.

Stiefel, V. L., D. C. Margolies & P. J. Bramel-Cox. 1992. Leaf temperature affects resistance to the Banks grass mite (Acari: Tetranychidae) on drought-resistant grain sorghum. J. Econ. Entomol. 85: 2170-2184.

Thindwa, H. P. & G. L. Teetes. 1994. Effect of temperature and photoperiod on sorghum resistance to biotype C and E greenbug (Homoptera: Aphididae). J. Econ. Entomol. 87: 1366-1372.

Tingey, W. M. & S. R. Singh. 1980. Environmental factors influencing the magnitude and expression of resistance, *In* F. G. Maxwell & P. R. Jennings, eds. *Breeding Plants Resistant to Insects,* pp 87-113. John Wiley & Sons, New York.

Tyler, J. M. & J. H. Hatchett. 1983. Temperature influence on the expression of resistance to Hessian fly (Diptera:Cecidomyiidae) in wheat derived from *Triticum tauschii.* J. Econ. Entomol. 76: 323-326.

van Houten, Y. M., W. P. Overmeer, A. Q. van Zon & A. Veerman. 1988. Thermoperiodic induction of diapause in the predacious mite *Amblyseiulus potentillae.* J. Insect Physiol. 34: 285-290.

van Steenis, M. J. 1994. Intrinsic rate of increase of *Lysiphlebus testaceipes* Cresson (Hym.: Braconidae), a parasitoid of *Aphis gossypii* Glover (Hom.; Aphididae) at different temperatures. J. Appl. Entomol. 118: 399-406.

Wadleigh, R. W., P. G. Koehler, H. K. Preisler, R. S. Patterson & J. L. Robertson. 1991. Effect of temperature on the toxicities of ten pyrethroids to German cockroach (Dictyoptera: Blattellidae). J. Econ. Entomol. 84: 1433-1436.

Walters, D. S., J. Harman, R. Craig & R. O. Mumma. 1991. Effect of temperature on glandular trichome exudate composition and pest resistance in geraniums. Entomol. Exp. Appl. 60: 61-69.

Wood, E. A., Jr. & K. J. Starks. 1971. Effect of temperature and host plant interaction on the biology of three biotypes of the greenbug. Envir. Entomol. 1: 230-234.

Wyatt, I. J. & S. J. Brown. 1977. The influence of light intensity, daylength and temperature on increase rates of four glasshouse aphids. J. Appl. Ecol. 14: 391-399.

Zareh, N. & J. G. Morse. 1989. Influence of temperature and life stage in monitoring for pesticide resistance in citrus thrips (Thysanoptera: Thripidae) with residual bioassays. J. Econ. Entomol. 82: 342-346.

6

Stored Product Integrated Pest Management With Extreme Temperatures

Linda J. Mason and C. Allen Strait

Contamination of stored grain with insects, insect fragments, fungi, and mycotoxins is a major concern of the grain industry. In the past, pesticides have been used to achieve both preventive and therapeutic control. However, insects are developing resistance to pesticides, customers are concerned about mycotoxins and pesticide residues, pesticides are being withdrawn from use because of high costs to re-register or develop new ones, and environmental considerations have made the manufacture of certain pesticides illegal. Therefore, alternative, preferably non-residual, methods of effective pest control are needed for the post-harvest handling, storage, and processing of grains, oilseeds, fruits, and vegetables.

The aim of this chapter is to examine how the use of extreme temperatures might benefit grain producers, handlers, and processors seeking to use non-chemical methods for preventing or controlling insect infestation and fungal development. Existing and potential technologies that might be considered include dielectric heating, microwaves, chilled aeration, fluidized beds, and heat sterilization. The mode of action, effectiveness, practical implementation, limitations, possible integration into current storage and handling facilities, and possible areas of future research are examined.

The Grain Ecosystem

A stored grain bulk is a man-made ecological system in which deterioration is an ongoing process, resulting from interactions among physical, chemical, and biological variables. Damage by insects, fungi, and sprouting causes hundreds of millions of dollars of economic losses to grain producers, merchandisers, and processors each year (Harein & Meronuck 1995). Although quality of harvested grain can never by improved with storage time, the rate of deterioration can be slowed with an integrated post-harvest management system that combines engineering, biological, and economic principles.

Alternative pest control techniques discussed within this chapter must focus on limiting pest reproduction and growth while managing quality factors. With the limited tools available, temperature is probably the most widely accepted treatment for grain conditioning. It is only recently that we have returned to the use of extreme temperatures as a pest management tool. In the long term, aeration cooling and other temperature modifying techniques may replace chemicals as the primary pest control method (Hagstrum & Flinn 1992). Before proceeding with specific uses of temperature, a brief review of the organisms we are attempting to control is essential.

Fungi Development and Mycotoxins

The presence of mold in grain can lead to kernel discoloration or moldy odors, resulting in economic losses associated with a decrease in grain quality. In addition, certain species of *Aspergillus, Fusarium,* and *Penicillium* can produce mycotoxins, which can make the grain unfit for human consumption, as well as cause detrimental effects when fed to animals (Wilson & Abramson 1992). Mold growth may also increase the grain temperature, thereby not only creating a more favorable environment for both increased insect and mold growth, but potentially allowing temperatures to rise high enough to scorch or even ignite the grain. Various fungi species require different conditions in which to survive and flourish. These conditions for a variety of fungal species are summarized in Table 6.1.

Certain *Aspergillus* and *Penicillum* species are the most important fungi species involved in the deterioration of stored grain, although at moisture contents above 22% *Fusarium, Alternaria, Epicoccum,* and *Mucor* species may be of importance (Stroshine et al. 1984). In addition to physical damage due to grain deterioration, some fungi can produce mycotoxins that can further reduce grain value. Mycotoxins are toxic substances produced by fungi (molds) growing on grain, feed, or food in the field or storage. Mycotoxins can be detrimental to the health of both animals and humans.

Although thousands of molds are capable of growing on stored grain, only a few mold species produce mycotoxins, the most important being *Aspergillus, Penicillum,* and *Fusarium.* The mycotoxins of concern are aflatoxin, deoxynivalenol (DON or vomitoxin), zearalenone, and fumonisin. All four of these are associated with ear rot disease in the field. However, mycotoxins can be produced in storage. This situation is almost always associated with improper drying and/or storage conditions.

Insects

Insects within stored feed and food cause numerous quality and health issues. Because of this, in the United States the Food and Drug Administration (FDA) and the Federal Grain Inspection Service (FGIS) set tolerances and grade standards regulating the number of insects found within raw and processed feed and food. The presence of insects and insect fragments above specified tolerances make the product illegal for human consumption. If FDA discovers lower than established levels of food quality, it has the authority to engage in a number of actions, including recall, injunction, or seizure. Raw grain that is found to have higher insect counts than the allowable standards is subject to rejection from trade or costly discounts to the seller. However, the highest standards are set by consumers. They generally have zero tolerance for insects in their food, and thus this is generally the final standard that must be met in a control strategy. Fortunately, insects and molds are sensitive to temperature changes within their environment and the postharvest environment is usually one that is enclosed, allowing for the manipulation of temperature. Thus, the use of temperature to restrict pest population is an excellent tool for the stored food industry.

Three basic temperature zones have been defined for insects and fungi: optimum, suboptimum, and lethal (Table 6.2). The optimum temperature zone is where the organism has the most rapid development and reproduction. Suboptimum temperatures are those either above or below the optimum where the insects and fungi are less fit, but are still able to survive and reproduce. Although not useful for killing insects, suboptimum temperatures are effective in managing insect populations in bulk stored grain by reducing population growth. Lethal temperatures are those above or below the suboptimum which will eventually kill the organism.

TABLE 6.1 Approximate temperature and relative humidity requirements for spore germination and growth of fungi common on corn kernels (Stroshine et al. 1984, reprinted with permission, Copyright, Purdue Research Foundation, West Lafayette, IN 47907)

Fungus	Minimum Relative Humidity (%)	Equilibrium Moisture Content (%)[1]	Growth Temperature (°C)		
			Minimum	Optimum	Maximum
Alternaria	91[2]	19	-4	20	36-40
Aspergillus glaucus	70-72	13.5-14	8	24	38
A. flavus	82	16-17	6-8	36-38	44-46
A. fumigatus	82	16-17	12	40-42	55
Cephalosporium acremonium[3]	97	22	8	24	40
Cladosporium	88	18	-5	24-25	30-32
Epicoccum	91	19	-4	24	28
Fusarium moniliforme	91	19	4	28	36
F. graminearum, F. roseum, (Giberella zeae)	94	20-21	4	24	32
Mucor	91	19	-4	28	36
Nigrospora oryzae[3]	91	19	4	28	32
Penicillium funiculosum[3]	91	19	8	30	36
P. oxalicum[3]	86	17-18	8	30	36
P. brevicompactum	81[2]	16	-2	23	30
P. cyclopium	81[2]	16	-2	23	30

| *P. viridicatum* | 81[2] | 16 | –2 | 23 | 36 |

[1] Approximate equilibrium moisture content at 25° C equal to the minimum percentage relative humidity at which fungus can germinate, probably below the moisture content that the fungus would be able to compete with other fungi on grain except for *A. glaucus*. The latter has no real competitor at 72% RH except occasionally *A. restrictus*.

[2] Approximately 5% or more of the population can germinate at this RH.

[3] Rarely found growing in stored grain regardless of moisture and temperature.

TABLE 6.2. The response of stored-product insects to temperature[1] (Reprinted from Journal of Stored Product Research, Vol. 28, P. G. Fields, The control of stored product insects and mites with extreme temperatures, pp. 89-118, Copyright (1992), with kind permission from Elsevier Sciences Ltd., The Boulevard, Langford Lane, Kidlington OX5 1GB, UK)

Zone	Temp. (°C)	Effect
Lethal	50-60	death in minutes
	45	death in hours
Suboptimum	35	development stops
	33-35	development slows
Optimum	25-33	maximum rate of development
Suboptimum	13-25	development slows
	13-20	development stops
Lethal	5	death in days (unacclimated) movement stops
	-10 to 5	death in weeks to months (acclimated)
	-25 to -15	death in minutes, insects freeze

[1]Species, stage of development and moisture content of food will influence the response

Low Temperature Strategies

Low suboptimum temperatures can slow the rate of damage from mold and insects, and therefore increase the length of time commodities can be safely stored (Burges & Burrell 1964, Howe 1965, Longstaff 1981, Mullen & Arbogast 1984, Brown & Hill 1984, Flinn & Hagstrum 1990). They might not, however, eliminate the infestation, or prevent further damage from occurring. In particular, molds are capable of growth below the freezing point, and can maintain populations in feed at temperatures as low as -23°C (Brown & Hill 1984). Fortunately insects have a narrower band of suboptimum temperatures. Generally, grain held between 13°C and 20°C will slow, but not stop, the development of most stored product insects.

The susceptibility of insects to lethal cold temperatures varies greatly between species and life stages, and is often dependent upon factors such as temperature, length of exposure, sex (Edwards 1958, Williams 1954, Le Torc'h 1977, Kawamoto et al. 1989), and relative humidity (Nagel & Shepard 1934, Birch 1953, Howe 1965, Stojanovic 1965, Le Torc'h 1977, Jacob &

Fleming 1986, Evans 1983, 1987b, Fields 1990). In general, species that are most easily controlled by cold temperatures include *Tribolium castaneum, T. confusum,* and *Oryzaephilus mercator* (Howe 1965, Sinha & Watters 1985), whereas *Trogoderma granarium, Sitophilus granarius, Ephestia kuehniella* and *Plodia interpunctella* are more difficult to control. Duration of exposure required to control various species at a given temperature was summarized by Fields (1992).

The most susceptible life stage is typically the egg (Watters 1966, Cline 1970, Daumal et al. 1974, Jacob & Fleming 1986, Johnson & Wofford 1991), which can be killed by exposure to 10°C for two weeks, or 9 h at -10°C (Watters 1966, Mullen & Arbogast 1979). In some species either the larvae or larvae and adults are the most cold tolerant stage (David et al. 1977).

Acclimation time appears to be very important in increasing the survival of insects at low temperatures (Evans 1983, Smith 1970, 1975, Fields 1990, Khan 1990). In general, exposing insects to cool temperatures prior to exposure to extreme temperatures increases their survival 2-10 times (Fields 1992). Thus, a more gradual cooling rate may require a longer treatment time to achieve the same mortality as could be obtained at a higher cooling rate in a shorter time.

Suboptimal low temperatures have been used widely to manage pest populations in bulk stored grain. This is especially true in the temperate regions of the world. Grain grown in the northern United States and Canada can be effectively cooled to safe storage temperatures (below 15°C) shortly after fall harvest. With additional cooling, bulk grain temperatures can be reduced to temperatures below -20°C (Burrell 1967, 1974, Sinha et al. 1979).

There are three basic ways in which grain storage temperatures can be lowered: turning, ambient or forced-air aeration, and chilled aeration.

Turning

Warm grain can be turned or mixed with cooler grain to help maintain a uniform temperature (Bryan & Elvidge 1977, Wilkin 1975, Muir et al. 1977). Turning can also achieve some insect control (Watters 1991). However, there are costs associated with turning grain for temperature control besides the labor and energy required to turn the grain. Any time grain is moved, including turning, there can be a loss of quality due to breakage. Additionally, there is little effect on the overall average grain temperature. However, if there are areas within the bulk which have heated, these "hot spots" are eliminated. "Hot spots" can be formed when insects or fungi grow, generate heat, and raise the local grain temperatures. Temperatures of 42-50°C have been recorded within hot spots (Banks & Fields 1995). This increased temperature allows for increased fungi and insect growth in the surrounding grain. Dissipation of these "hot spots" is important in reducing the rate of grain deterioration.

Ambient Aeration

A second method for lowering grain temperatures is ambient aeration (Burrell 1967, Calderon 1974, Bhatnagar & Bakshi 1975, Armitage & Burrell 1978, Calderwood et al. 1984, DeJean 1992). Aeration involves the use of fans to ventilate grain storage facilities. This process helps to maintain the grain at a uniform temperature, usually within a few degrees of the ambient temperature, and can sometimes manage "hot spots". Maintenance of near ambient temperatures helps prevent convection currents, which can lead to moisture migration, and therefore crusting, mold, and insect growth (Foster & McKenzie 1979). Unfortunately, the ability to maintain low enough temperatures is weather dependent. In some regions there may not be enough cool days or nights for the grain to be sufficiently cooled, while in the northern regions of the world, near freezing temperatures can easily be achieved throughout most of the storage period.

Studies conducted in the mid-western United States from June 1 to September 14, 1994 concluded that, due to the fact that the ambient temperature was not below the target of 15°C for long enough time periods, maintenance of grain temperature below 15°C would have been impossible with ambient aeration (Maier et al. 1996). Thus, insect control and grain conditioning through the use of ambient aeration alone are not feasible in some geographic areas or climates.

In temperate climates, grain managers can slow insect population growth by aeration with cool, ambient air (Cuperus et al. 1986, 1990). During fall harvest, grain managers can utilize aeration fans during cool days or nights to remove the heat of drying or harvest. In general, cooling grain to temperatures below 5°C arrests development and prevents population growth until the following spring. However, grain stored in tropical or temperate climates that is either harvested during the summer months (wheat and rice) or stored over the summer until the next harvest season (food maize and popcorn), often cannot be cooled by aeration fans due to high ambient temperatures.

Chilled Aeration

A third method involves the use of refrigerated air instead of ambient air (Sutherland et al. 1970, Navarro et al. 1973, Elder 1984, Cunney et al. 1986, Mason et al. 1994, Maier 1992, 1994). Ambient air is cooled by refrigeration coils and then reheated a few degrees to decrease the relative humidity. The major advantage of chilled over ambient aeration is the ability to control the temperature and relative humidity of the airflow independent of ambient weather conditions. Therefore it is possible not only to cool the grain more rapidly, but also to a desired temperature below that which would be possible through conventional aeration. Due to the insulating properties of stored grain, once the target temperature has been reached, only short periods of rechilling are required to maintain the desired temperature. The lack of a need for continual treatment allows for the use of mobile chilling units that can be moved to multiple storage facilities as required.

Tests of various chilling units have been conducted throughout the world, including Great Britain (Burrell 1967, Burrell & Laundon 1967), Israel (Donahaye et al. 1974), Australia (Hunter & Tayor 1980, Thorpe & Elder 1980) and the United States (Maier et al. 1996, 1997, Mason et al. 1997). Comparisons between conventional ambient aeration and chilled aeration of stored popcorn in the midwestern United States during the summer of 1994 have demonstrated several advantages of chilled aeration (Maier et al. 1997, Mason et al. 1997). In contrast to ambient aeration, chilled aeration was able to maintain the grain temperature near the target of 15°C, while grain temperatures for conventional aeration were 6-18°C higher (Fig. 6.1). Grain chilling was also more effective in controlling insects than ambient aeration in combination with chemical control. Higher numbers of hairy fungus beetle, *Typhaea stercorea*, and Indianmeal moth were detected in the traditional treatment of aeration plus fumigation as compared to the chilled aeration treatment (Figs. 6.2, 6.3). Finally, rapid cooling of grain during periods of warm ambient temperatures should also prevent cold-acclimation, thus achieving greater mortality than ambient fall cooling.

Although there are several advantages of chilled aeration, major disadvantages are the high capital and running costs (Bell & Armitage 1992). However, recent economic comparisons indicate that cost of refrigeration of stored grain can be competitive with conventional pest management expenses (Table 6.3). Cost of ambient aeration plus two fumigations is approximately 0.096-0.17 cents/kg popcorn. This is comparable with the expense of chilled aeration, which is 0.11 cents/kg popcorn (Mason et al. 1997).

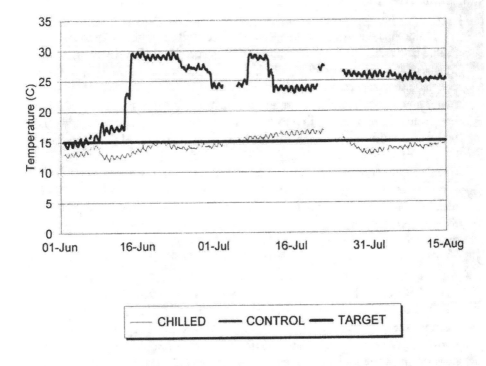

FIGURE 6.1 Overall mean popcorn temperatures measured in control bins versus chilled bins during the 1994 summer storage period (Maier et al. 1997). Reprinted with kind permission of Elsevier Sciences, Ltd., The Boulevard, Langford Lane, Kidlington 0X5 1GB, United Kingdom.

Freezing

Although typically not recommended, temperatures $< 0\,^\circ C$ can be used on bulk grain, provided moisture levels of the air are not high enough to wet the grain. If frozen grain is unloaded in warm, humid weather, condensed moisture can encourage mold growth (McKenzie & Van Fossen 1980). In smaller quantities of bulk food and certainly when attempting to control pantry pests, temperatures $< 0\,^\circ C$ are easily and safely achieved. Generally, 4 d at $-18\,^\circ C$ will kill most pantry pests in home stored food. It is important to note that the extreme temperatures must be reached within the center of the bulk food, or pests will find a warm refuge and not be controlled. Temperature recommendations for some bulk products are listed in Table 6.4.

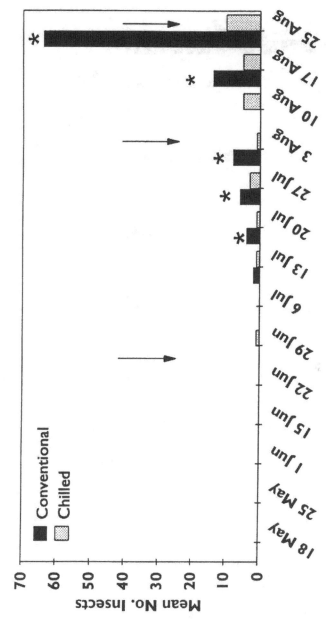

FIGURE 6.2 Mean number of hairy fungus beetles, *Typhaea stercorea*, collected in grain probe traps (9 traps/bin) of conventional (ambient aeration and fumigation) and chilled aeration bins (Mason et al. 1997). Bars, by date, with asterisks are significantly different at the P<0.05 level of significance. Arrows indicate fumigation. (Reprinted with kind permission from Elsevier Sciences Ltd, The Boulevard, Langford Lane, Kidlington 0X5 1GB, UK)

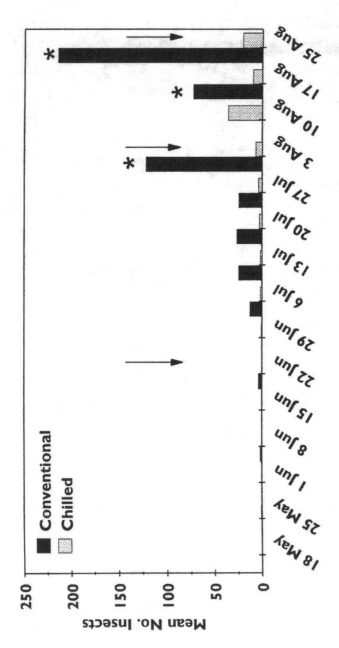

FIGURE 6.3 Mean number of Indianmeal moths, *Plodia interpuntella*, collected in grain probe traps (9 traps/bin) of conventional (ambient aeration and fumigation) and chilled aeration bins (Mason et al. 1997). Bars, by date, with asterisks are significantly different at the P<0.05 level of significance. Arrows indicate fumigations. (Reprinted with kind permission from Elsevier Sciences Ltd, The Boulevard, Langford Lane, Kidlington 0X5 1GB, UK)

TABLE 6.3. Economic comparison of chilled vs. ambient aeration treatments (Mason et al. 1997)

Cost Category	Chilled Aeration	Ambient Aeration
Treatment cost (cents/hr)	42	28
Treatment cost (cents/kg popcorn)	0.11	0.023
Fumigation cost (cents/kg popcorn)	NA	0.037-0.073
Number of fumigations per summer	0	2
Total cost (cents/kg popcorn)	0.11	0.097-0.169

High Temperature Strategies

Like low temperatures, exposure to temperatures only 5°C above the optimum are capable of slowing or stopping insect activity and development, and, depending on the species, are capable of causing death. Exposures to temperatures between 42-50°C for short periods of time (seconds to a few hours) generally produce over 90% mortality (Table 6.5). The ability of an insect to survive high temperatures is dependent upon the same factors that

TABLE 6.4. Chilling times for selected commodities exposed in a 0.76 m² (27 ft³) freezer filled to capacity (Mullen & Arbogast 1979, reprinted with permission)

Commodity	Freeze setting (0°C)	Time to 0°C (h)	Time to equilibrium (h)
Cornflakes	-10	7	30
(28-3.2 lb cases)	-15	6	30
	-20	5	35
Flour	-10	55	160
(7-100 lb bags)	-15	29	130
	-20	25	145
Elbow macaroni (15-24	-10	29	130
lb cases) & Blackeyed	-15	18	95
peas (15-24 cases)	-20	19	100

TABLE 6.5 Temperatures, exposure times, and mortalities for various stored product insects exposed to high temperatures

Order Species	Stage	Peak Temp. (°C)	Time (sec)	% Mortality	Heating System	Reference
Coleoptera						
Lasioderma serricorne	Adult	49	20	35	Infrared	Kirkpatrick & Tilton (1972)
	Adult	57	32	99	Infrared	Kirkpatrick & Tilton (1972)
	Adult	66	40	100	Infrared	Kirkpatrick & Tilton (1972)
Oryzaephilus surinamensis	Egg	81	120	99	Microwave	Locatelli & Traversa (1989)
	Larva, pupa	80	120	99	Microwave	Locatelli & Traversa (1989)
	Adult	35	2.2	50	High frequency	Nelson & Kantack (1966)
	Adult	44	4.4	90	High frequency	Nelson & Kantack (1966)
	Adult	49	20	97	Infrared	Kirkpatrick & Tilton (1972)
	Adult	57	32	98	Infrared	Kirkpatrick & Tilton (1972)
	Adult	66	40	100	Infrared	Kirkpatrick & Tilton (1972)
Rhyzopertha dominica	Egg, larva, pupa	55	10	50	Infrared	Tilton & Schroeder (1963)
	Egg, larva, pupa	63	15	10	Infrared	Tilton & Schroeder (1963)
	Egg, larva, pupa	64	16	100	Fluidized bed	Fleurat-Lessard (1985)
	Egg, larva, pupa	68	25	100	Fluidized bed	Evans (1987a)
	Egg, larva, pupa	82	120	100	Microwave	Locatella & Traversa (1989)
	Egg, larva, pupa	56	290	53	Fluidized bed	Evans & Dermott (1981)
	Egg, larva, pupa, adult	65	–	99	Fluidized bed	Vardel & Tilton (1981)
	Egg, larva, pupa, adult	67	290	100	Fluidized bed	Vardel & Tilton (1981)
	Adult	49	20	45	Infrared	Kirkpatrick & Tilton (1972)
	Adult	57	32	97	Infrared	Kirkpatrick & Tilton (1972)

Sitophilus granarius	Adult	66	40	100	Infrared	Kirkpatrick & Tilton (1972)
	Adult	32	3	40	Microwave	Baker et al. (1956)
	Adult	42	6	55	Microwave	Baker et al. (1956)
	Adult	57	9	98	Microwave	Baker et al. (1956)
	Adult	72	12	100	Microwave	Baker et al. (1956)
Tribolium castaneum	Adult	49	20	62	Infrared	Kirkpatrick & Tilton (1972)
	Adult	57	32	98	Infrared	Kirkpatrick & Tilton (1972)
	Adult	66	40	100	Infrared	Kirkpatrick & Tilton (1972)
Lepidoptera						
Ephestia cautella	Pupa	40	7,200	0	Incubator	Arbogast (1981)
	Pupa	45	7,200	6	Incubator	Arbogast (1981)
	Pupa	45	7,200	100	Incubator	Arbogast (1981)
Plodia interpunctella	Pupa	40	7,200	3	Incubator	Arbogast (1981)
	Pupa	45	7,200	100	Incubator	Arbogast (1981)
	Pupa	50	7,200	100	Incubator	Arbogast (1981)

influence low temperature survival: relative humidity, acclimation, exposure duration, species, and developmental stage (Howe 1965, Fields 1992).

A major difference between high and low temperature treatments is the length of exposure necessary for control. Lethal cold temperatures usually require longer exposure times than is necessary at lethal high temperatures (Fields 1992). However, prolonged exposure to high temperatures may increase the risk of damage to the commodity being treated, including germination and baking quality (Lindberg & Sorenson 1959, Kreyger 1972, Ghaly & Taylor 1982). Therefore, when high temperature treatments are applied, it is usually advantageous to quickly recool the commodity. A slow recool may keep grain excessively long in the optimum temperature zone for infestation or mold growth to reoccur. Recooling of the product can also cause excessive moisture loss (shrink) which ultimately can cause significant economic loss.

High temperature treatments for bulk stored products can be applied several different ways: fluidized and sprouted heated beds (60-120°C), high frequency electric fields (approximately 10-100 MHZ), microwaves (approximately 0.5-3 GHz) and infrared (approximately 100-100,000 GHz.). Most methods attempt to heat the grain to temperatures above 60°C for brief periods of time (seconds) (Kirkpatrick & Tilton 1972, Locatelli & Traversa 1989, Fleurat-Lessard 1985, Tilton & Schroeder 1963, Tilton et al. 1983). Determination of which method to use will ultimately depend on local utility rates, construction and operating costs, as well as public acceptance of the method. The major drawback to all systems is the rate at which bulk grain can be treated at normal grain transfer rates (tons per hour). Only one method, fluidized bed heating, has been attempted at a full scale prototype facility. All other methods remain at pilot stage or laboratory scale levels (Thorpe 1983, Sutherland et al. 1986, Banks & Fields 1995).

Fluidized and Sprouted Beds

Fluidized beds utilize hot air (60-120°C) blown at rates high enough to lift, mix, and quickly heat the grain (Dermott & Evans 1978, Fleurat-Lessard 1985). This can be accomplished with either a continuous flow system, where grain is constantly moving through the heater, or a batch system in which one load is treated at a time (Dermott & Evans 1978). Trials on the lesser grain borer, *Rhyzopertha dominica*, one of the most heat-tolerant stored product insects, determined that the lethal times required to kill 99.9% of the population ($LT_{99.9}$) for 1 kg samples at 80° and 140°C were 222 and 43 sec., respectively (Evans & Dermott 1981). The relationship between inlet air temperature and mortality is nonlinear and is not dependent upon the final grain temperature alone. This indicates that the heating rate is also an important factor (Evans 1987a, b).

If large-scale treatments of grain are required, the operating costs of fluidized beds are similar to the cost of fumigation treatments. Costs for batch treatments of grain were comparable to fumigation with methyl bromide (Dermott & Evans 1978). The first continuous flow pilot plant was constructed in Victoria, Australia in 1982 (Evans et al. 1983). A combination of control features was utilized including entoleters, preheating chambers, fluidized beds (65-70°C; airflow rates of 21.-2.4 kg/sec/m²; 2.5-4.5 min. duration), and evaporative cooling. Complete insect mortality was achieved without any loss of grain quality (Evans et al. 1984).

Sprouted beds, designed in Canada for grain drying, can be used for crops such as maize that are difficult to manipulate in fluidized beds (Mathur & Gishler 1955). In a sprouted bed, the air enters through a nozzle above which the grain is fluidized. The grain then circulates, falls into the surrounding area of low velocity air, then reenters the airstream. Inlet temperatures generally are 80-180°C. Results indicated that a bed 1 m in diameter could disinfest 3-4 tons per hour, with larger diameter beds increasing outputs to almost 15 metric tons per hour. An added benefit of sprouted beds over fluidized beds is that insects are killed by both mechanical damage as well as heat stress. Damage to grain in a sprouted bed, despite circulation, is negligible (Mathur & Gishler 1955).

Dielectric Heating

Another utilization of extreme temperatures which shows promise for stored products is to take advantage of the principle of dielectric heating. When exposed to high frequency electrical fields both the insects and products will be rapidly heated as they absorb the electrical energy. However, because the rate of energy absorption depends upon the electrical characteristics of the exposed material, it is possible to heat the insects to a higher temperature more quickly than the surrounding grain. This differential rate of dielectric heating can enable the destruction of the insects without damage to the grain. The critical electrical properties (dielectric constant and dielectric loss factor) are frequency and temperature dependent (Nelson 1986).

The practical implementation of dielectric heating would be facilitated with knowledge of the electrical characteristics of several stored product pests and host mediums. This knowledge base could potentially improve the efficiency of dielectric heating by optimizing the frequency for various pest-host combinations. This would also require identification of pest species prior to treatment. For example, the frequency necessary for effective control of larval Indianmeal moths infesting corn may be different than the frequency required for control of adult flour beetles infesting wheat; however, the differing frequencies could be determined for each specific case prior to treatment. The most effective frequency could then be utilized. Once the variation with frequency is known, the temperature dependence could possibly be utilized for improved efficiency by modifying the frequency during the treatment in order to maintain the maximum heating differential between pest and host (Nelson & Stetson 1974).

The three factors which must be considered when examining electric field control are frequency, field intensity, and heating rate (Nelson 1995). The most effective frequencies fall in the range of 10-100 MHz (Nelson & Stetson 1974). At these frequencies the electrical properties of the pest and host are such that the difference in energy absorption rate is greatest. In general, with intensities above 1200 kV/m, there is little improvement in effectiveness with increasing field intensity. Higher heating rates are desired to minimize the loss of heat to the surroundings, particularly during longer treatments. However, the final temperature, and not the rate of heating, is most important in determining insect mortality (Nelson 1995).

Frequency dependence of the dielectric constant and dielectric loss factor has been measured for hard red winter wheat, *Triticum aestivum*, and rice weevils, *Sitophilus oryzae* (Nelson & Charity 1972). Dielectric constants decreased continuously with increasing frequency throughout the range from 250 Hz to 12.2 GHz, while dielectric loss factors decreased with increasing frequency from 250 Hz to a minimum in the region of 50 KHz, then increased to a peak in the region between 5 and 100 MHz. The highest insect-to-grain loss-factor ratios were noted in the frequency range of 100 KHz and 1 GHz. Based on the existing literature Nelson & Charity (1972) concluded that the most promising region for selectively heating insects was 10 to 100 MHz.

The susceptibility of insects to the heating effects of electric fields varies with species and developmental stage and has been extensively documented (Webber et al. 1946, Baker et al. 1956, Bollaerts et al. 1956, Van den Bruel et al. 1960, Nelson & Whitney 1960, Whitney et al. 1961, Van Dyck 1965, Nelson & Kantack 1966, Nelson et al. 1966, Nelson 1973, Benz 1975, Anglade et al. 1979). In general, adults are more susceptible than the larval stage, although this susceptibility varies among species. A summary of the susceptibility of various life stages of three species of stored grain beetles is presented in Table 6.6. These differences may be due to specific biological or physiological characteristics, as well as the surrounding medium, size variations, and the geometry of the insect. Another factor is the host-medium particle size. Adult rice weevils treated in 5 mm glass beads experienced higher mortality than those treated in 0.84 mm glass beads (Nelson et al. 1966). The kernel may also offer some protection. Insect mortality was found to be higher for adult rice weevils and lesser grain borer positioned outside the kernel, as compared to those same species, stage, and age treated inside the kernel (Nelson et al. 1966).

Some of the requirements for differential dielectric heating include: the high frequency electrical conductivity of the pest must exceed that of the host, the applied voltage must be adequate, and the smallest linear dimension of the pest must be approximately 0.04 cm for an unmodulated field (Thomas 1952). A pulse modulated field may be effective in instances where the above conditions are not met. For example, due to the small size of fungi and bacteria, a pulse modulated field may be necessary. However, pulse-modulated fields have not been found to improve insect control efficiency (Nelson 1995).

In addition to mortality, exposure to an electric field may also cause physical injuries and reduced reproduction rates in surviving insects. Physical injuries most often occur in the appendages, particularly the leg joints. Treated larvae may develop into adults with deformed or missing legs (Nelson 1995). Although surviving insects were capable of reproduction, the more severe treatment levels reduced the rate of reproduction. Lesser grain borers exhibited lower reproduction rates when exposed to treatments resulting in >50% mortality (Nelson et al. 1966). Both the physical injuries and reduced reproduction were attributed to thermal damage. No non-thermal effects attributable to electric fields have been found (Nelson 1995).

A large portion of the expense of using high frequency electric fields for the control of stored product insects involves the capital cost of the equipment. The rate of movement of the grain through the equipment is also an important consideration. Because the field intensity is inversely related to the distance between the electrodes across which the field is applied, the grain must be passed through the equipment in shallow layers. This would either require slower grain handling rates, or larger equipment than currently exists (Nelson 1995). Due to the significant capital equipment costs the most practical utilization of dielectric heating would be permanent sites at mills and elevators with a large grain handling capacity, or export facilities which handle large

TABLE 6.6 Reported host-media temperatures following radio frequency and microwave dielectric heating exposures necessary for 99-100% mortality of several stored product beetles (Nelson et al. 1995, reprinted with permission).

Species	Stage	Frequency (MHz)	Medium	Temperature[1] (°C)	Reference
Sitophilus oryzae	Mixed immature	27	Wheat	56	Anglade et al. (1979)
	Adult	39	Wheat	39	Nelson & Whitney (1960)
	Mixed immature	39	Wheat	61	Nelson & Whitney (1960)
	Adult	39	Wheat	40	Nelson & Stetson (1974)
	Adult	2,450	Wheat	83	Nelson & Stetson (1974)
	Adult	2,450	Wheat	>60	Tateya & Takano (1977)
	Pupa	2,450	Wheat	>60	Tateya & Takano (1977)
	Larva	2,450	Wheat	>58	Tateya & Takano (1977)
	Egg	2,450	Wheat	>57	Tateya & Takano (1977)
	Adult	2,450	Wheat	65	Hamid & Boulanger (1969)
	Egg	2,450	Flour	>57	Tateya & Takano (1977)
	Larva	2,450	Flour	>58	Tateya & Takano (1977)
	Pupa	2,450	Flour	>60	Tateya & Takano (1977)
	Adult	2,450	Flour	>66	Tateya & Takano (1977)
	Larva	2,450	Flour	>82	Baker et al. (1956)
	Adult	2,450	Flour	>68	Baker et al. (1956)
	Larva	90	Flour	53	Van den Bruel et al. (1960)
Sitophilus granarius	All	13.6	Wheat	62	Benz (1975)
	Egg	13.6	Wheat	61	Benz (1975)
	Adult	27	Wheat	55	Anglade et al. (1979)
	Larva	27	Wheat	58	Anglade et al. (1979)

	Pupa	27	Wheat	61	Anglade et al. (1979)
	Adult	39	Wheat	41	Nelson & Kantack (1966)
	Adult	39	Wheat	42	Nelson et al. (1966)
	Adult	2,450	Wheat	86	Anglade et al. (1979)
	Adult	2,450	Wheat	>92	Hamid et al. (1968)
	Adult	2,450	Wheat	>57	Baker et al. (1956)
	Larva	2,450	Wheat	>82	Baker et al. (1956)
	Egg	2,450	Wheat	>72	Baker et al. (1956)
Tribolium confusum	Adult	11	Flour	75	Webber et al. (1946)
	Larva	11	Flour	65	Webber et al. (1946)
	Adult	27	Flour	60	Anglade et al. (1979)
	Larva	27	Flour	65	Anglade et al. (1979)
	Adult	39	Wheat	47	Nelson & Whitney (1960)
	Adult	39	Wheat shorts	>60	Nelson & Whitney (1960)
	Adult	90	Flour	59	Van den Bruel et al. (1960)

[1]> symbol indicates that 99-100% mortality not obtained by indicated temperature; temperatures reported by Tateya & Takeno (1977) are for 95% mortality.

quantities of grains. Another disadvantage is that an electric field would have no residual effect upon the grain, which would leave it susceptible to immediate reinfestation. The lack of residual protection would also make it necessary to recool the grain to acceptable storage temperatures to slow the growth of any insect reinfestations or mold. Product heating and recooling will also cause additional water loss (shrink) in the grain, a feature representing additional economic costs to the grain handler.

Due to the high cost of electric field application, estimated at $0.07/bu (Nelson 1995), it is not currently practical for use in the control of stored product pests. Unless alternate control factors such as reduced life span, reduced reproduction rate, or behavior changes can be associated with non-thermal effects of electric field exposure, the application of electric field dielectric heating is not likely a viable cost effective alternative for current control methods (Nelson 1995).

Microwaves

Microwaves are very similar to high frequency electric fields, but utilize much higher frequencies, 0.5-3 GHz. This enables rapid heating (72-83°C) to be achieved with much lower field intensities. However, the penetration depth becomes more important, as the microwave intensity diminishes with increased penetration. Also, the electrical properties of the pest and host are much closer at these frequencies, which makes differential heating less effective. Unlike electric field frequencies (10-100 MHz), the grain is heated close to the same temperature as the insects. The effectiveness also depends upon the moisture content of the grain. A higher moisture content yields higher temperatures and therefore higher mortality for a given treatment time (Baker et al. 1956, Watters 1976, Locatelli & Traversa 1989). As the grain is heated, the moisture content of the grain will drop, reducing the rate of temperature increase, and potentially reducing mortality. This is unlikely to be a factor due to the short treatment times (seconds) (Baker et al. 1956, Locatelli & Traversa 1989). One hundred percent mortality of confused flour beetle in wheat at 15.6% m.c. was obtained after a 105 second exposure to 8.5 GHz with a pulse frequency of 4 MHz (Watters 1976).

As frequency increases grain temperatures must increase to generate complete mortality. In a comparison between a 39 MHz electric field and a 2.45 GHz microwave on the mortality of rice weevils in wheat, the 39 MHz frequency achieved complete mortality at 40°C wheat temperatures, compared to 80°C for complete mortality at 2.45 GHz (Nelson & Stetson 1974). This was accomplished through the differential heating of the insect and grain. At 39 MHz, the insects absorb the energy at a greater rate than the wheat. Therefore the temperature of the individual insects within the grain will be greater than the grain bulk, i.e. while the grain bulk is at 40°C, the insects within the grain may be at 80°C. Because the rate of energy absorption is similar for the insects and grain at 2.45 GHz, both are heated at the same rate, and would therefore reach the same temperature of 80°C. This accounts for the difference in bulk grain temperatures for the different treatments.

Research in the field of ultra-high frequencies ($\geq$10 GHz) indicates that these frequencies are effective in obtaining selective heating of insect pests (Halverson et al. 1996). Tests at 10.6 GHz indicate that while this frequency is less effective (less heating differential between host grain and pest insect resulting in higher bulk grain temperatures) than 39 MHz treatments, the heating differential is greater at 10.6 GHz than at 2.45 GHz (Halverson et al. 1996). Therefore, depending upon how the dielectric properties vary with frequency, it is possible frequencies above 10.6 GHz could prove to be more efficient than even the frequencies in the 10-100 MHz range (Fig. 6.4). From the 10-100 MHz range, the potential for differential heating drops through the 2.45 GHz range, and then increases again at the 10.6 GHz range. Because dielectric heating properties are unknown past 10.6 GHz, it cannot be said with certainty whether the potential for dielectric heating will increase or decrease. If the dielectric properties are such that differential heating increases beyond 10.6 GHz, then these frequencies will offer the most efficient and effective control. If the differential heating decreases again beyond 10.6 GHz, then frequencies within the 10-100 MHz will remain most effective.

Microwaves are limited in application much the same way as high frequency electric fields. Due to limited penetration the grain would have to be treated in shallow layers. There would also be no residual effects, and the grain would become susceptible to immediate reinfestation. An advantage with microwaves is the lack of interference with television and radio waves, which could potentially be a problem at the lower radio frequencies (Thomas 1952). Estimates for the cost of microwave treatment at an energy cost of $0.05/kW-hr range from $2.05 per metric ton of wheat to $2.74 per metric ton of ground wheat (Halverson et al. 1996).

In addition to effects on insects, microwaves have been shown to destroy fungi on sorghum grain. A treatment time of 60 seconds at 1.25 GHz was lethal to most fungal spores/mycelium found on sorghum, independent of moisture content or microwave power (Table 6.7).

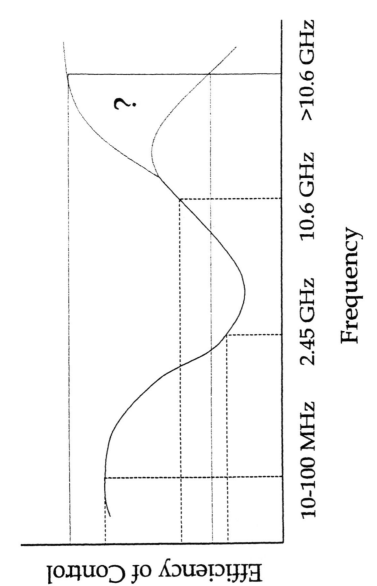

FIGURE 6.4 Qualitative comparsion of the efficiency of insect control in a bulk commodity (based upon lowest host medium temperature) for various radio frequencies

Heat Sterilization

Heat sterilization involves heating an entire processing plant or warehouse to a target temperature for several hours (Sheppard 1984). The temperature is achieved and maintained through the use of the facility's heating system, as well as additional heating units. The entire facility is shut down during the treatment, and can therefore result in economic losses due to downtime in addition to treatment costs. Therefore, it is desirable to reduce the treatment time as much as possible. This may be possible through combination with other treatment methods, such as traditional fumigation and an oxygen deficient atmosphere. A guideline for the required times to bring about insect mortality are presented in Table 6.8. This method is currently receiving much attention from industries limited in the use of traditional chemicals, such as food processing plants. As more pesticides are removed from use within a food production facility, food plants are looking for economical, safe ways to eliminate pests. However, more information is needed on the long-term effects on pest populations and equipment before heat sterilization can become more widely adopted.

Combination Techniques/Controlled Atmospheres

An extensive research base exists for controlled atmospheres (Calderon & Barkai-Golan 1990); the topic is included in this chapter because it combines favorably with extreme temperatures. Controlled atmospheres usually involve increasing the concentration of nitrogen or carbon dioxide gas to obtain oxygen deficient and/or carbon dioxide enriched conditions within a storage structure. An advantage is that they can easily be combined with other methods of pest control, such as heat sterilization and fumigation. In general, the reduction of oxygen in the environment stresses the insect, thereby making it more susceptible to other control methods, including non-chemical as well as traditional methods. An example of this is the combination of carbon dioxide, high temperatures, and low concentrations of phosphine. At CO_2 levels between 6-19%, temperature of approximately $30°C$, and phosphine concentrations of around 165 ppm (reduced from the normal 850-1500 ppm), complete mortality of all stages of red flour beetle, rice weevil, Angoumois grain moth, and warehouse beetle, was obtained after a 20 h treatment period (Mueller 1995).

TABLE 6.7 The effect of microwave treatment levels and time on percentage isolation of fungi from sorghum grain. Data are means of 50 plated sorghum seeds. (Reprinted from Journal of Stored Product Research, Vol 28, More, Magan, & Stenning, Effects of microwave heating on quality and mycoflora of sorghum grain, Pages No. 251-256. Copyright (1992), with kind permission from Elsevier Sciences Ltd, The Boulevard, Langford Lane, Kidlington OX5 1GB, UK)

			Percentage Fungi Isolated from Grain at Power Level (kW)								
			0.15			0.3			0.6		
			Percentage Moisture			Percentage Moisture			Percentage Moisture		
Fungus	Time (sec.)	Control	12	14	16	12	14	16	12	14	16
Acremonium spp.	30	10	14	18	0	6	0	0	0	0	0
	60	10	2	0	0	0	0	0	2	0	0
Alternaria spp.	30	4	2	0	0	2	0	0	2	0	0
	60	4	2	0	0	2	0	0	0	0	0
Aspergillus candidus	30	50	40	28	0	26	8	0	0	0	0
	60	50	4	0	0	4	0	0	0	0	0
A. flavus	30	4	4	0	0	0	0	0	0	0	0
	60	4	4	0	0	0	0	0	0	0	0
A. niger	30	26	2	2	0	8	10	2	0	0	0
	60	26	4	0	0	10	0	0	2	0	0
Cladosporium spp.	30	6	6	6	0	4	4	0	2	0	0
	60	6	4	0	0	2	0	0	0	0	0
Eurotium spp.	30	70	78	74	52	18	4	6	6	8	0
	60	70	0	0	0	0	0	0	0	0	0

Penicillium spp.	30	30	84	10	0	16	10	0	6	6	0
	60	30	32	0	0	2	0	0	2	0	0
Rhizopus spp.	30	12	12	6	0	4	4	4	2	6	0
	60	12	6	0	0	2	0	0	0	0	0

TABLE 6.8 Generalized stored product insect lethal times at various temperatures (Fields 1992)

Temperature (°C)	Time
40	24 h
45	12 h
50	5 min
55	1 min
60	30 sec

Summary

With increasing consumer concerns over the potential health and environmental hazards of pesticides, as well as insect resistance to and loss of these, non-chemical control methods will become increasingly important in future pest management strategies. Although it is unlikely that a "magic bullet" will be developed, there is potential for the integration of extreme temperatures into current pest management practices, whether as replacements for, or companions to, current chemical treatments. Many different temperature based technologies may have application in various niches of post harvest storage, handling, and processing, with potential for combination with other non-temperature based technologies. For example, grain from the field may be disinfected by electric fields, microwaves, or fluidized beds. Temperature treated grain may then be cooled through chilled aeration to prevent insect growth throughout the grain bulk. Whole grain may be processed in a plant disinfested with a carbon dioxide/heat sterilization treatment. Applications which provide continued control, such as ambient aeration, and chilled aeration can easily be combined with quick, non-residual disinfestation procedures such as electric fields, microwaves, and various heat treatments in an integrated pest management program.

There are many different combinations which could be employed, and more research needs to be conducted to determine which treatments are most effective and economical in various applications and combinations. Electric fields and microwaves would require greater efficiencies to cover the increased capital and maintenance costs incurred or consumer and environmental demands for chemical-free products to compete as cost-effective alternatives to current control methods.

Practical aspects of application are also important in determining the feasibility of these technologies. For example, the use of electric fields for the attraction of insects may have limited effectiveness due to interference from the many man-made electric field sources in the environment, such as power lines. The biological effects of electric and magnetic fields are not completely understood, and potential health concerns for operators of electric field and microwave equipment must be considered.

Public perception is possibly the most important factor in determining the value of various applications. For example, microwave technology has been widely accepted by the general populace, and would most likely be easy to market as a safe alternative to chemicals. Whatever new technologies are developed, it will be important to ensure public acceptance, compliance with environmental regulations, and the development of proper safety procedures, in addition to achieving acceptable, cost-effective control.

Given this review of the literature with respect to past and present research efforts on the suitability of temperature based technologies for the control of insects and fungi in stored crops, the following conclusions can be drawn:

(1) Electric fields in the range of 10 to 100 MHz are effective when used for the dielectric heating of insects in stored grains. Other applications of electromagnetic fields, such as attraction/repulsion, physical injury, and disruption of reproduction due to non-thermal effects, show little promise.

(2) Microwaves in the frequency range of 2.45 to 10.6 GHz require higher grain temperatures than temperatures required in the range of 11 to 90 MHz to achieve complete mortality. Application of microwaves is limited due to insufficient penetration depth.

(3) Low temperatures can slow insect and mold development, thereby increasing storage time. Mortality in insects is dependent upon cooling rate as well as final temperature. However, molds are able to grow at sub-freezing temperatures.

(4) High temperatures applied in either fluidized or sprouted beds, can quickly cause destruction of insects and molds. Treatment time is dependent upon heating rate and final temperature. The treated commodity requires recooling to slow reinfestation and product quality must not be altered.

(5) Controlled atmospheres involving oxygen deficient environments can improve the effectiveness and efficiency of traditional and alternative pest control measures when combined with them.

Recommendations for Future Research Directions

Based on this review of the literature and the current state of the art of extreme temperatures applied to stored product insect pest management, the following specific recommendations for future research directions are made: (1) The effect of long term exposure to electric fields on insects should be further examined. This will allow for the determination of either the absence or presence of any non-thermal effects. Such data could also provide insight as to the effects of electric fields upon non-arthropods as well as the environment. The ability of insects to detect and react to electric fields should also be examined. Emphasis should be placed upon the determination of the physical ability for the insect to detect electric fields. Destruction of insects through dielectric heating may be cost effective when dealing with high value items, such as rare artifacts or high value food items. The dielectric properties for a variety of insect species have not yet been determined.
(2) The application of chilled aeration technologies to various crops and effects of environmental factors should be further studied.
(3) Heat sterilization and fluidized bed technology should be further examined to determine the most efficient ways of incorporating them into current pest management strategies. The effect of high temperatures upon the commodity and equipment must be known to evaluate any potential costs due to product loss (shrink) or machinery depreciation. A procedure for recooling of the commodity while preserving its quality must also be established.
(4) The combination of oxygen deficient atmospheres with modified temperature environments should be examined further. The combination of controlled atmosphere with heat sterilization in particular may reduce treatment time or enable the use of lower temperatures.

Acknowledgments

This is manuscript no. AES 15490, Purdue University Agricultural Research Program.

References

Anglade, P. H. Cangardel & F. Fleurat-Lessard. 1979. Application de O.E.M. de hate frequence et des micro-ondex a la desinsectisation des denrees stockees. *Microwave Power Symp. 1979 Digest* (XIV symposium International sur les Applications Energetiques des Micro-ondes, Monace, 11-15 juin 1979), pp 67-69.

Arbogast, R. L. 1981. Mortality and reproduction of *Esphestia cautella* and *Plodia interpunctella* exposed as pupae to high temperatures. Environ. Entomol. 10: 708-710.

Armitage, D. M. & N. J. Burrell. 1978. The use of aeration spears for cooling infested grain. J. Stored Prod. Res. 14: 223-226.

Baker, V. H., D. E. Wiant & O. Taboada. 1956. Some effects of microwaves on certain insects which infest wheat and flour. J. Econ. Entomol. 49: 33-37.

Banks, J. & P. Fields. 1995. Physical methods for insect control in stored-grain ecosystems, *In* D. S. Jayas, N. D. G. White & W. E. Muir, eds. *Stored-Grain Ecosystems*, pp 353-409, Marcel Dekker, Inc., New York.

Bell, C. H. & D. M. Armitage. 1992. Alternative storage practices, *In* D. B. Saver, ed. *Storage of Cereal Grain and Their Products*, pp 249-340. Amer. Assoc. Cereal Chem., Inc., St. Paul, Minnesota.

Benz, G. 1975. Entomologisch Untersuchungen zur Entwesung von Getreide mittels Hochfrequenz. Alimenta 14 (1): 11-15.

Bhatnagar, A. P. & A. S. Bakshi. 1975. Aeration studies on the storage of wheat grains in a 50-tonne outdoor metal bin. Punjab Agric. Univ. J. Res. 12: 189-199.

Birch, L. C. 1953. Experimental background to the study of the distribution and abundance of insects. I. The influence of temperature, moisture and food on the innate capacity for increase of three grain beetles. Ecology 34: 698-711.

Bollaerts, D., F. Pietermaat & W. E. van den Bruel. 1956. Destruction des déprédateurs dans les aliments par les champes électriques á hautes frequence. *C.R. 8e Symp. Phytopharmacie*, pp 449-458, Gand.

Brown, C.W. & S. T. Hill. 1984. Survival of micro-organisms in deep-frozen barley and pig feed. J. Stored Prod. Res. 20: 145-150.

Bryan, J. M. & J. Elvidge. 1977. Mortality of adult grain beetles in sample delivery systems used in terminal grain elevators. Can. Entomol. 109: 209-213.

Burges, H. D. & N. J. Burrell. 1964. Cooling bulk grain in the British climate to control storage insects and to improve keeping quality. J. Sci. Food Agric. 15: 32-50.

Burrell, N. J. 1967. Grain cooling studies - II: effect of aeration on infested grain bulks, J. Stored Prod. Res. 3: 145-154.

Burrell, N. J. 1974. Chilling, *In* C. M. Christensen, ed. *Storage of Cereal Grains and Their Products.*, Am. Assoc. Cereal Chem., St. Paul, Minnesota.

Burrell, N. J. & J. H. Laundon. 1967. Grain cooling studies - I: Observations during a large scale refrigeration test on damp grain. J. Stored Prod. Res. 3: 125-144.

Calderon, M. 1974. Aeration of grain-benefits and limitations. OEPP/EPPO Bull. 6: 83-94.

Calderon, M. & R. Barkai-Golan, eds. 1990. *Food Preservation by Modified Atmospheres*. CRC Press, Boca Raton, Florida. 402 pp.

Calderwood, D. L., R. R. Cogburn, B. D. Webb & M. A. Marchetti. 1984. Aeration of rough rice in long term storage. Trans. Amer. Soc. Agric. Engin. 27: 1579-1585.

Cline, D. L. 1970. Indian-meal moth egg hatch and subsequent larval survival after short exposures to low temperature. J. Econ. Entomol. 63: 1081-1083.

Cunney, M. B., P. A. Williams & S. J. Morley. 1986. Application of engine driven heat pumps to grain drying with refrigerated storage. Comm. Eur. Comm. Rep. EUR 10303. 137 pp.

Cuperus, G. W., R. T. Noyes, W. S. Fargo, B. L. Clary, D. A. Arnold & K. Anderson. 1990. Management practices in a high-risk, stored wheat system in Oklahoma. Amer. Entomol. 36: 129-134.

Cuperus, G. W., C. K. Prickett, P. D. Bloome & T. J. Pitts. 1986. Insect populations in aerated and unaerated wheat in Oklahoma. J. Kan. Entomol. Soc. 59: 620-627.

Daumal, J., P. Jourdheuil & R. Tomassone. 1974. Variabilite des effets letaux des basses temperatures en fonction du stade de developpement embryonnaire chex la pyrale de la farine (*Anagasta kuhniella* Zell., Lepid., Pyralidae). Ann. Zool.-Ecol. Anim. 6: 229-243.

David, H. H., R. B. Mills & G. D. White. 1977. Effects of low temperature acclimation on developmental stages of stored-product insects. Environ. Entomol. 6: 181-184.

DeJean, J. 1992. Continental Grain Company, Chicago Division. Marketing and Maintenance of Quality in Market Channels, NC-151 Annual Meeting. 12 February 1992, Chicago, Illinois.

Dermott, T. & D.E. Evans. 1978. An evaluation of fluidized bed heating as a means to disinfesting wheat. J. Stored Prod. Res. 14: 1-12.

Donahaye, E., S. Navarro & M. Calderon. 1974. Studies on aeration with refrigerated air - III. Chilling of wheat with a modified chilling unit. J. Stored Prod. Res. 10: 1-8.

Edwards, D. K. 1958. Effects of acclimatization and sex on respiration and thermal resistance in *Tribolium* (Coleoptera: Tenebrionidae). Can. J. Zool. 36: 363-382.

Elder, W. B. 1984. Aeration with naturally occurring and refrigerated air, *In* B. R. Champ & E. Highley, eds. *Proc. Aust. Dev. Asst. Course on Preservation of Stored Cereals* (1981). CSIRO. Canberra, Australia.

Evans, D. E. 1983. The influence of relative humidity and thermal acclimation on the survival of adult grain beetles in cooled grain. J. Stored Prod. Res. 19: 173-180.

Evans, D. E. 1987a. The influence of rate of heating on the mortality of *Rhyzopertha dominica* (F.) (Coleoptera: Bostrychidae). J. Stored Prod. Res. 23: 73-77.

Evans, D. E. 1987b. The survival of immature grain beetles at low temperatures. J. Stored Prod. Res. 23: 79-83.

Evans, D. E. & T. Dermott. 1981. Dosage-mortality relationships for *Rhyzopertha dominica* (F.) (Coleoptera: Bostrychidae) exposed to heat in a fluidized-bed. J. Stored Prod. Res. 17: 53-64.

Evans, D. E., G. R. Thrope & T. Dermott. 1983. The disinfestation of wheat in a continuous-flow fluidized bed. J. Stored Prod. Res. 19: 125-137.

Evans, D. E., G. R. Thrope & J. W. Sutherland. 1984. Large scale evaluation of fluid bedheating as a means of disinfesting grain, *In 3rd Proc. Int. Work. Conf. Stored Prod. Entomol.* pp 523-530.

Fields, P. G. 1990. The cold-hardiness of *Cryptolestes ferrugineus* and the use of ice-nucleation active bacteria as a cold-synergist, *In* F. Fleurat-Lessard & P. Ducom, eds. *Proc. 5th Int. Work. Conf. Stored-Prod. Prot.,* pp 1183-1191. Bordeaux, France.

Fields, P. G. 1992. The control of stored-product insects and mites with extreme temperatures. J. Stored Prod. Res. 28: 89-118.

Fleurat-Lessard, F. 1985. Les traitements thermiques de desinfestation des cereales et des produits cerealiers: Possibilite d'utilisation pratique et domain d'application. Bull. OEPP. 15: 109-118.

Flinn, P. W. & D. W. Hagstrum. 1990. Simulations comparing the effectiveness of various stored-product management practices used to control *Rhyzopertha dominica* (Coleoptera: Bostrichidae). Environ. Entomol. 19: 725-729.

Foster, G. H. & B. A. McKenzie. 1979. Managing grain for year-round storage. AE-90. Dept. Agric. Engin. Coop. Exten. Serv. Purdue Univ., W. Lafayette, Indiana. 5pp.

Ghaly, T. F. & P. D. Taylor. 1982. Quality effects of heat treatment of two wheat varieties. J. Agric. Eng. Res. 27: 227-234.

Hagstrum, D. W. & P. W. Flinn. 1992. Integrated pest management of stored-grain insects, *In* D. B. Sauer, ed. *Storage of Cereal Grains and Their Products,* pp 535-562. Amer. Assoc. Cereal Chem., St. Paul, Minnesota.

Halverson, S. L., W. E. Burkholder, T. S. Bigelo, E. V. Nordheim & M. E. Misenheimer. 1996. High-power microwave radiation as an alternative insect control method for stored products. J. Econ. Entomol. 89: 1638-1648.

Hamid, M. A. K., C. S. Kashyap & R. Van Cauwenberghe. 1968. Control of grain insects by microwave power. J. Microwave Power. 3: 126-135.

Hamid, M. A. K. & R. J. Boulanger. 1969. A new method for the control of moisture and insect infestations of grain by microwave power. J. Microwave Power 4: 11-18.

Harein, P. & R. Meronuck. 1995. Stored grain losses due to insects and molds and the importance of proper grain management, *In* V. Krischik, G. Cuperus & D. Galliart, eds. *Stored Product Management,* pp 29-31. E-912. CES. Div. Agric. Sci. Nat. Res. OSU. USDA. FGIS. ES. APHIS. 242 pp.

Howe, R. W. 1965. A summary of estimates of optimal and minimal conditions for population increase of some stored products insects. J. Stored Prod. Res. 1: 177-184.

Hunter, A. J. & P. A. Taylor. 1980. Refrigeration aeration for the preservation of bulk grain. J. Stored Prod. Res. 16: 123-131.

Jacob, T. A. & D. A. Fleming. 1986. The effect of temporary exposure to low temperature on the viability of eggs of *Oryzaephilus surinamensis* (L.) (Col., Silvanidae). Ent. Month. Mag. 122: 117-120.

Johnson, J. A. & P. L. Wofford. 1991. Effects of age on response of eggs of Indianmeal moth and navel orangeworm (Lepidoptera: Pyralidae) to subfreezing temperatures. J. Econ. Entomol. 84: 202-205.

Kawamoto, H., R. N. Sinha & W. E. Muir. 1989. Effect of temperature on adult survival and potential fecundity of the rusty grain beetle, *Cryptolestes ferrugineus*. Ann. Ent. Zool. 24: 418-423.

Khan, N. I. 1990. The effects of various temperature regimes and cooling rates on the mortality and reproductive abilities of two stored grain insect species. M.Sc. Thesis, Oklahoma State University.

Kirkpatrick, R. L. & E. W. Tilton. 1972. Infrared radiation to control adult stored-product Coleoptera. J. Ga. Ent. Soc. 7: 73-75.

Kreyger, J. 1972. Drying and storing grains, seed, and pulses in temperate climates. Publikatie 205. Inst. Bewer, Verwek Landb. Prod. IBVL. 333 pp.

Le Torc'h, J. M. 1977. Le froid: moyen de protection contre les ravageurs des denrees stockees. Essais de laboratoire sur les insectes de Pruneaux. Revue Zool. Agricole Path. Veg. 76: 109-117.

Lindberg, J. E. & E. I. Sorenson. 1959. Relationship between critical kernel temperature and moisture contents with respect to germinating properties of wheat. Kunfl. Skogsoch. Lantbruks. Akademiens Tidsk. Suppl. 1.

Locatelli, D. P. & S. Traversa. 1989. Microwaves in the control of rice infestations. Ital. J. Food Sci. 2: 53-62.

Longstaff, B. C. 1981. The manipulation of the population growth of a pest species: an analytical approach. J. Appl. Ecol. 18: 727-736.

MacKenzie, B. A. & L. D. Van Fossen. 1980. Managing dry grain in storage. AED-20. Midwest Plan Serv. Iowa State Univ.

Maier, D. E. 1992. The chilled aeration and storage of cereal grains. Ph.D. thesis, Michigan State University, East Lansing.

Maier, D. E. 1994. Chilled aeration and storage of U.S. crops - a review, *In* E. Highley, E. J. Wright, H. J. Banks & B. R. Champ, eds. *Proc. 6th Int. Working Conf. Stored Prod. Prot.*, pp 300-311. Canberra, Australia, 1994. Vol. 1.

Maier, D. E., W. H. Adams, J. E. Throne & L. J. Mason. 1996. Temperature management of the maize weevil, *Sitophilus zeamais* Motsch. (Coleoptera: Curculionidae) in three locations in the United States. J. Stored Prod. Res. 32: 255-273.

Maier, D. E., R. A. Rulon & L. J. Mason. 1997. Chilled versus ambient aeration and fumigation of stored popcorn - Part 1: Temperature management. J. Stored Prod. Res. 28: 33-39.

Mason, L. J., R. A. Rulon & D. E. Maier. 1997. Chilled versus ambient aeration and fumigation of stored popcorn - Part 2. Pest management. J. Stored Prod. Res. 28: 41-58.

Mason, L. J., D. E. Maier, W. H. Adams & J. L. Obermeyer. 1994. Pest management of stored maize using chilled aeration - a mid-west United States perspective, *In* E. Highley, E. J. Wright, H. J. Banks & B. R. Champ, eds. *Proc. 6th Int. Working Conf. Stored-Prod. Prot.*, pp 312-317. Canberra, Australia, 1994. Vol. 1.

Mathur, K. & P. Gishler. 1955. A study of the application of the sprouted bed technique to wheat drying. J. Appl. Chem. 5: 624.

More, H. G., N. Magan & B. C. Stenning. 1992. Effect of microwave heating on quality and mycroflora of sorghum grain. J. Stored Prod. Res. 28: 251-256.

Mueller, D. K. 1995. A new method of using low levels of phosphine in combination with heat and carbon dioxide. Fumigation Service and Supply, Inc. Indianapolis, Indiana. Handout. 4 pp.

Mullen, M. A. & R. T. Arbogast. 1979. Time-temperature-mortality relationship for various stored-product insect eggs and chilling times for selected commodities. J. Econ. Entomol. 72: 476-478.

Mullen, M. A. & R. T. Arbogast. 1984. Low temperatures to control stored product insects, *In* F.J. Baur, ed. *Insect Management for Food Storage and Processing*, pp 257-264. Amer. Assoc. Cereal Chem., St. Paul, Minnesota.

Muir, W. E., G. Yaciuk & R. N. Sinha. 1977. Effects on temperature and insect and mite populations of turning and transferring farm-stored wheat. Can. Agric. Engin. 19: 25-28.

Nagel, R. H. & H. H. Shepard. 1934. The lethal effect of low temperatures on the various stages of the confused flour beetle. J. Agric. Res. 48: 1009-1016.

Navarro, S., E. Donahaye & M. Calderon. 1973. Studies on aeration with refrigerated air - I. Chilling of wheat in a concrete elevator. J. Stored Prod. Res. 9: 253-259.

Nelson, S. O. 1973. Insect-control studies with microwaves and other radiofrequency energy. Bull. Entomol. Soc. Amer. 19: 157-163.

Nelson, S. O. 1986. Potential agricultural application for RF and microwave energy. Paper 86-6539. Amer. Soc. Agric. Engin., St. Joseph, Michigan.

Nelson, S. O. 1995. Assessment of RF and microwave electric energy for stored-grain insect control, *Annual Internat. ASAE Meeting*, June 18-23, 1995, Chicago, Illinois. Amer. Soc. Agric. Engin., St. Joseph, Michigan. 16 pp.

Nelson, S. O. & B. H. Kantack. 1966. Stored-grain insect control studies with radio-frequency energy. J. Econ. Entomol. 59: 588-594.

Nelson, S. O. & L. F. Charity. 1972. Frequency dependence of energy absorption by insects and grain in electric fields. Trans. Amer. Soc. Agric. Engin. 15: 1099-1102.

Nelson, S. O. & L. E. Stetson. 1974. Comparative effectiveness of 39- and 2450-MHz electric fields for control of rice weevils in wheat. J. Econ. Entomol. 67: 592-595.

Nelson, S. O. & W. K. Whitney. 1960. Radio-frequency electric fields for stored grain insect control. Trans. Amer. Soc. Agric. Engin. 3: 132-137.

Nelson, S. O., L. E. Stetson & J. J. Rhine. 1966. Factors influencing effectiveness of radio-frequency electric fields for stored-grain insect control. Trans. Amer. Soc. Agric. Engin. 809-815, 817.

Sheppard, K. O. 1984. Heat sterilization (superheating) as a control for stored-grain pests in a food plant, *In* F.J. Baur, ed. *Insect Management for Food Storage and Processing,* pp 194-200. Amer. Assoc. Cereal Chem., St. Paul, Minn.

Sinha, H., H. A. H. Wallace, J. T. Mills & R. I. H. McKenzie. 1979. Storability of farm-stored hulless oats in Manitoba. Can. J. Plant Sci. 59: 949-957.

Sinha, R. N. & F. L. Watters. 1985. *Insect Pests of Flour Mills, Grain Elevators and Feed Mills and Their Control.* Agric. Canada Publ. 1776E, Canadian Govt. Publ. Centre, Ottawa, Canada. 290 pp.

Smith, L. B. 1970. Effects of cold-acclimation on supercooling and survival of the rusty grain beetle, *Cryptolestes ferrugineus* (Stephens) (Coleoptera: Cucujidae) at sub-zero temperatures. Can. J. Zool. 48: 853-858.

Smith, L. B. 1975. The role of low temperatures to control stored food pests. *Proc. 1st Int. Working Conf. Stored-Product Entomol.,* pp 418-430. Savannah, Georgia, 1974.

Stojanovic, T. 1965. Uticaj via nosti psenice na otpornost I aka (*Calandra granaria* L. I Cl oryzae L.) I itnog kukulji ara (*Rhyzopertha dominica* F.) prema niskm temperaturama. Let. Nauch. Radova Poljoprivr. Fakult. Novom Sadv 9: 80-90.

Stroshine, R., J. Tuite, G. H. Foster & K. Baker. 1984. *Self-Study Guide for Grain Drying and Storage.* Purdue Research Foundation. Purdue Univ., W. Lafayette, Indiana, 131 pp.

Sutherland, J. W., G. R. Thorpe & P. W. Fricke. 1986. Grain disinfestation by heating in a pneumatic conveyor, *In Proc. Cong. Agric. Engin. Adelaide,* pp 419-425. Canberra, Institution of Engineers, Melbourne, Australia.

Sutherland, J. W., D. Postcode & H. J. Griffiths. 1970. Refrigeration of bulk stored wheat. Aust. Refrig. Air Condit. Heat. 24: 30-34, 43-45.

Tateya, A. & T. Takano. 1977. Effects of microwave radiation on two species of stored product insects. Res. Bul. Plant Prot. Japan 14: 52-59.

Thomas, A.M. 1952. Pest control by high-frequency electric fields: critical résumé. Technical Report W/T 23. British Electrical and Allied Industries Res. Assoc., Leatherhead, Surrey, England. 40 pp.

Thorpe, G. E. 1983. High temperature disinfestation of grain. CSIRO (Australia) Research Review. 1983: 41-47.

Thorpe, G. E. & W. B. Elder. 1980. The use of mechanical refrigeration to improve the storage of pesticide treated grain. Internat. J. Refrig. 3: 99-106.

Tilton, E. W. & H. W. Schroeder. 1963. Some effects of infrared irradiation on the mortality of immature insects in kernels of rough rice. J. Econ. Entomol. 56: 727-730.

Tilton, E. W., H. H. Vardell & R. D. Jones. 1983. Infrared heating with vacuum for control of the lesser grain borer (*Rhyzopertha dominica* F.) and rice weevil (*Sitophilus oryzae* L.) infesting wheat. J. Georgia Entomol. Soc. 18: 61-64.

Van den Bruel, W. E., D. Bollaerts, F. Pietermaat & W. Van Dijck. 1960. Etude des facteurs determinant les possibilitiés d'utiisation du chauffage diélectrique á haute fréquence pour la destruction des insectes et des acariens dissimules en frofondeur dans les denrees alimentaires empaquetees. Prasitica 16: 29-61.

Van Dyck, W. 1965. La destruction des insects et des acariens dans les grains et a la farine qumoyen d'un champ électrique á haute fréquence. Rev. Agric. 18: 4455-462.

Vardell, H. H. & E. W. Tilton. 1981. Control of the lesser grain borer, *Rhyzopertha dominica* (F.), and the rice weevil, *Stitophilus oryzae* (L.), in wheat with a heat fluidized bed. J. Kans. Entomol. Soc. 54: 481-485.

Watters, F. L. 1966. The effects of short exposures to sub-threshold temperatures on subsequent hatching and development of eggs of *Tribolium confusum* Duval (Coleoptera, Tenebrionidae). J. Stored Prod. Res. 2: 81-90.

Watters, F. L. 1976. Microwave radiation for control of *Tribolium confusum* in wheat and flour. J. Stored Prod. Res. 12: 19-25.

Watters, F. L. 1991. Physical methods to manage stored-food pests, *In* J. R. Gorham, ed. *Ecology and Management of Food-Industry Pests*, pp 399-413. Assoc. Official Anal. Chem., Arlington, Virginia.

Webber, H. H., R. P. Wagner & A. G. Pearson. 1946. Higher-frequency electric fields as lethal agents for insects. J. Econ. Entomol. 39: 487-498.

Whitney, W. K., S. O. Nelson & H. H. Walkden. 1961. Effects of high-frequency electric fields on certain species of stored-grain insects. Market Res. Rept. 455, MQRD, AMS, USDA.

Wilkin, D. R. 1975. The effects of mechanical handling and the admixture of acaricides on mites in farm-stored barley. J. Stored Prod. Res. 11: 87-95.

Williams, G. C. 1954. Observations of the effect of exposure to a low temperature on *Laemophloeus minutus* (Ol.) (Col., Cucujidae). Bull. Entomol. Res. 45: 351-359.

Wilson, D. M. & D. Abramson. 1992. Mycotoxins, *In* D. B. Saver, ed. *Storage of Cereal Grain and Their Products*, pp 341-391. Amer. Assoc. Cereal Chem., St. Paul, Minnesota.

7

Use of Extreme Temperatures in Urban Insect Pest Management

Michael K. Rust and Donald A. Reierson

Extreme temperatures may be used in some situations to control insects of urban importance. Beetles, moths, and termites have been controlled for decades by extreme temperatures in sensitive places such as museums and food storage facilities. In this chapter we discuss the use of low and high temperatures to control insect pests that attack buildings or which may be found in homes, apartments, museums, and similar sensitive places where an alternative control strategy may be warranted or in situations which preclude the use of insecticides.

While the concept of utilizing heat or cold to kill insects is not new, the technology to produce rapid extreme temperature change is new. Ultra-cold and blast refrigerators, bulk liquid nitrogen, high-efficiency propane heaters, and portable microwave units are but a few of the advances in recent years that make use of temperature extremes a practical and attractive control strategy. In this review, we highlight studies concerning the use of extreme temperatures to kill wood-destroying organisms and common domiciliary pests, especially pests that are of importance to consumers and the structural pest control industry.

Extreme Low Temperatures

Responses of insects to cold may be categorized as follows: (1) chill coma, -3 to $10°C$; (2) cold shock, 5 to $-17°C$; (3) freeze avoidance, -1 to $-65°C$; and (4) freeze tolerance, -2 to below $-70°C$ (adapted from Lee 1991). For the uninitiated, it may be surprising to learn that many kinds of insects can survive long periods of freezing. Insects are immobilized during chill coma, but the coma is reversible if the exposure is not too long. Cold shock or direct chilling injury results from a rapid drop in temperature and brief exposure to low but nonfreezing conditions. Some cellular or tissue membrane damage is likely and, if sufficiently severe, cold shock results in death (Denlinger et al. 1991). For example, the flesh fly *Sarcophaga crassipalpis* exposed to -10°C for 2 h continued to develop but died in the pharate adult stage just prior to eclosion (Chen et al. 1987). Death presumably occurred because of interference with physiological events that coordinate eclosion behavior and because of damage to various organ systems (Denlinger et al. 1991). Depending upon the insect, brief pre-exposures to 0-6°C may prevent cold shock injury by initiating protective physiological mechanisms. Using the same example of *S. crassipalpis*, the flies readily survived 2-h exposure at -10°C when they were first exposed to 0°C for 2 h (Chen et al. 1987). This phenomenon is fairly common among insects. Many kinds of insects survive exposure to low temperatures by utilizing a variety of physiological mechanisms including supercooling, ice nucleation, freezing, cold hardening, and cyroprotection (Lee 1991).

Even though Salt (1961) indicated that household pests typically do not encounter freezing temperatures and, therefore, would not be expected to have evolved mechanisms to tolerate freezing temperatures, some stages such as eggs of the cigarette beetle, *Lasioderma serricorne*, (Mullen & Arbogast 1984) and larvae of the hide beetle, *Dermestes maculatus*, (Strang 1992) readily survive freezing temperatures. Apparently, some cold-protective mechanisms have indeed evolved in these specific cases before they became household pests. In general, few studies have been made with urban pests to determine if they are capable of acclimating to low temperatures (Strang 1992).

In using extreme low temperature to kill urban insect pests, it is especially crucial that rate of cooling, the minimum lethal temperature, or the rate of thawing exceed the physiological limit or threshold of the insect to survive. Generally, a cooling rate of 1-5°C min^{-1} does not alter the supercooling point, the temperature at which body water spontaneously freezes, more than 1-2°C. For example, cooling rates of 0.17°C min^{-1} allow the northern tenebrionid beetle, *Upis ceramboides*, to survive low temperatures, whereas rates exceeding 0.32°C min^{-1} were lethal (Miller 1978). Low temperature acclimation usually accomplished by exposing insects to low temperatures typically triggers enhanced supercooling capacity (the ability to withstand successively lower temperatures) in freeze-tolerant insects (Lee 1991). Some organisms such as bacteria, spermatozoa, and yeast cells survive if they are pre-frozen on a strict temperature-lowering regimen and kept at -20 to -3 °C for a day. However, this kind of pre-freezing conditioning is not sufficient to protect higher animals such as insects when they are exposed to super-low temperatures (Asahina 1966). Cooling and thawing rates found in the insect's microhabitat are especially important for the ability of an insect to survive low temperature (Baust & Rojas 1985).

Household Pests and Insects in Museum Artifacts

Carpets, mattresses, linens, pillows, and upholstered furniture have been treated with liquid N_2 to control house dust mites, *Dermatophagoides* spp. (Shibasaki et al. 1996). Storage of furs or textiles at -10 °C to prevent feeding damage by carpet beetles and clothes moths has been recommended for nearly 100 years. However, because -10 °C does not kill all stages of these pests, repetitive cycles of thawing and chilling are recommended to eventually ensure that affected articles are completely disinfested. For example, all stages of clothes moths were killed when infested items were chilled to -7.8 °C for several days, warmed to 10 °C for a short time, returned to -7.8 °C, and then stored at about 4.5 °C (Back 1935). Recent measurements indicated that eggs of the webbing clothes moth, *Tineola bisselliella*, may survive at -20 °C for up to 30 h (Brokerhof et al. 1993). Larvae may survive storage at -7° to -4 °C for up to 67 days (Back & Cotton 1927).

Rapid freezing has been widely used to successfully control museum pests (Florian 1986, 1990, Gilberg & Brokerhof 1991, Strang 1992, 1995, Tanimura & Yamaguchi 1995). Smith (1984) recommends the use of a blast freezer to lower exposure temperatures quickly. Air in a blast freezer circulates over evaporator coils, accelerating the rate of freezing and attaining temperatures down to -40 °C so rapidly that insect cold-adaptive mechanisms are rendered irrelevant. Treatments in such a freezer are made in three-stages: cool down (12 h), exposure (24 h), and a gradual 12-h recovery to room temperature. In a modified version of this strategy, Nesheim (1984) eliminated an infestation of the anobiid beetle *Gastrallus* spp. in books by freezing books in plastic bags to -30 °C within 4-5 h in a blast freezer and holding the books at that temperature for 72 h. Pieces of aluminum foil inserted between about every 100 pages increased the rate of freezing of the books. Such foil inserts may be practical for assisting the rate of freezing of other objects as well. However, not all experts recommend that books be frozen in polyethylene bags (Smith 1984). When bags are used with books or other objects that are water-saturated, condensation forms in the bag, but it never forms from objects acclimatized to median relative humidities. If bags are to be used, objects should be allowed to dry before being bagged and frozen (Florian 1992). Absorbent cotton wool can be used to moderate changes in relative humidity in the bags (Kronkright 1992).

Freezing should be done with caution and should be avoided with some objects and under certain conditions. Not all objects are amenable to freezing. Wood objects held together with desiccated or partially failing glue, deteriorated finishes, and inlays or veneers showing mechanical damage should not be frozen. Expansion and contraction during the freezing or thawing process may damage the item or exacerbate damage that already exists. Objects that should never be frozen include paintings on canvas, ivory, ancient and deteriorating glass, high-fire ceramics, and waterlogged specimens (Kronkright 1992). If possible, small valueless samples of a like material should be subjected to the freezing regimen to ensure the object can satisfactorily withstand the rigors of freezing.

Careful handling of objects subjected to low temperatures is necessary. Polymeric materials become stiffer and more brittle at low temperatures. Drying oil films become glassy below $-30\,°C$, and acrylics become glassy below $0\,°C$ and leathery between 0 and $-50\,°C$ (Strang 1995). Changes in the properties of wood exposed to low temperatures are reversible (Florian 1986), whereas effects on adhesives vary. For instance, Kite (1992) reported that adhesives used to repair leather are not affected by freezing at $-30\,°C$ for 3 days; longer exposures were not tested. The use of cold to control insect pests in mammal collections has generally been discontinued because humidity, hydration and expansion-contraction due to extreme temperature damages hides and hair (Williams et al. 1985).

Because protection from cold provided to insects by brief exposures to $0\,°C$ is lost within 2 h after the insect is returned to $25\,°C$ (Chen et al. 1991), many museums alternate freezing and thawing to kill insect pests. Tanimura & Yamaguchi (1995) modified the procedure of Florian (1990) to effectively kill two species of carpet beetles with cold. Objects were wrapped in soft paper and placed in polyethylene bags. A vacuum was drawn on the bag, and it was sealed. The bagged items were held at $18\,°C$, then rapidly cooled and held at $-30\,°C$ for 48-72 h. The bag was then thawed for 1 day after which freezing at $-30\,°C$ was repeated for another 48-72 h. The bag was then thawed again for 1 day and the objects were left in the polyethylene bag to prevent reinfestation. Twelve of 34 curators responding to a survey use this method to kill insect pests in museums objects (Tanimura & Yamaguchi 1995).

In the literature, authors typically report only successful low temperature treatments. The actual minimum lethal temperature is rarely reported (Strang 1992). However, some general recommendations for cold treatments for museum artifacts are as follows: herbarium materials are successfully disinfested by exposures of -18°C for 48 h, insects in books at -20°C and -29°C for 72 h, and general museum pests at -29°C for 48 h (Florian 1986). Based on this information, it appears that exposures of -30°C for 72 h will disinfest virtually any museum object.

Wood-Destroying Organisms

Use of liquid N_2 to control localized infestations of drywood termites has been used commercially since 1986 (Darlin 1994). Forbes & Ebeling (1986) reported that western drywood termites, *Incisitermes minor*, succumbed from short exposures at about -20°C. Rapid injection of liquid N_2 along with good insulation of the area being cooled can lower wood temperatures to such an extent that termites cannot survive. Approximately 26 liters of liquid N_2 are piped into each void between studs. Because so much liquid N_2 is used to treat wall voids, a treatment may be quite costly. Treatment of a 6-meter exterior wall can cost about US $2,000.

Because the minimum temperatures required to kill several kinds of wood-destroying insects have not been reported, we exposed the drywood termites *I. minor* and *I. syderi*, Formosan subterranean termites, *Coptotermes formosanus*, and southern lyctus beetles, *Lyctus planicollis*, to freezing conditions to determine the critical thermal minimum (CT_{min}) for each of them. Various developmental stages of each species were confined in wooden blocks in an ultracold freezer, and temperature was critically measured with thermocouples. The core temperature of the blocks was lowered at $1°C$ min^{-1} until the blocks reached a desired low temperature. The blocks and insects were then removed and thawed for 24 h. Survivorship of the exposed insects is shown in Table 7.1. These experiments indicated that termites and wood-destroying beetles die if exposed for even a brief period to -17 to -30°C. Of course, they succumb even more quickly when exposed to significantly lower temperatures.

We conducted additional studies to investigate the potential of liquid N_2 to control structural insect pests. We used a mock-up non-bearing wall under laboratory conditions to determine if liquid N_2 could reduce temperatures in

TABLE 7.1 The critical thermal minimum temperatures (CT_{min}) for various wood-destroying insects in wood blocks (Rust, Reierson & Paine, unpublished data)[a]

Insect	Stage	Temperature °C (±95% CL) to kill		
		10%	50%	100%
Western drywood termite	workers	-4.3 (-12.7 to 4.9)	-11.8 (-20.6 to -3.4)	-21.3 (-31.9 to -13.0)
	alates	-10.4 (-15.1 to -4.1)	-15.3 (-20.2 to -9.8)	-21.4 (-27.4 to -16.0)
Southeastern drywood termite	workers	-8.9 (-12.7 to -5.1)	-12.6 (-15.7 to -8.9)	-17.2 (-22.4 to -13.2)
Formosan subterranean termite	workers	-8.1 (-11.2 to -5.5)	-14.1 (-19.3 to -7.1)	-16.5 (-21.0 to -8.2)
Southern lyctus beetle	eggs	-14.9 (-34.1 to -6.5)	-23.8 (-33.0 to -14.0)	-34.9 (-52.1 to -19.1)
	small larvae	-13.6 (-18.1 to -0.5)	-18.3 (-23.1 to -10.5)	-24.1 (-32.7 to -19.7)
	large larvae	-15.5 (-19.6 to -10.6)	-19.3 (-23.4 to -15.0)	-24.0 (-28.7 to -19.9)

[a] 10 insects per replicate (n=5) in a 0.5-dram vial in a chamber in Douglas fir block (14 cm x 10 cm x 5 cm) split longitudinally and held together with rubber bands. Blocks placed in an ultracold freezer at -50°C so that the temperature in the block decreased about 1°C min^{-1}.

a wall to insect CT_{min}'s (Rust et al. 1995). We presumed that insects die if the temperature of their immediate surroundings reaches their CT_{min}. The wall was built with construction grade Douglas fir [nominal 2 x 4-inch (4.4 x 8.3 cm) on 40.6 cm (16 inch) centers], was 2.44 m high, had nominal 2 x 4-inch top and base plates, and had removable fireblocks 96 cm from the floor plate. One side of the wall was permanently covered with 1.3 cm gypsum plaster wallboard and the other side was covered with removable ACX 1.3 cm plywood panels. Seam joints in the wallboard were taped and sealed with joint compound. Electronic bead-type flexible wire thermocouples placed in the center of wall studs at 11 different locations were used to determine the temperature in the wood. Effects on insects were determined from the resultant mortality among groups of insects in wood blocks exposed to the cold in strategic locations in the wall while it was treated.

Liquid N_2 was delivered to the wall from a commercial 110-liter Dewar via a flexible 1.2-m woven stainless steel hose. To determine how the point of introduction affected resultant temperatures, nitrogen was introduced near the top of individual wall voids as follows: (1) centered 10 cm below the top plate, (2) centered 112 cm below the plate (15 cm above the fireblock), and (3) centered 18 cm above the bottom plate. Rate of application was regulated by weighing the Dewar as the nitrogen was introduced. Thermocouples placed in the studs of adjacent voids help determine the lateral sphere of cooling. Although insulation mats on wall surfaces help achieve lower temperatures more quickly (Forbes & Ebeling 1986), we did not use insulation mats in our study. Therefore, our study provided a conservative measure of the cooling profile of a section of a wall treated with liquid N_2.

As proposed from previous exposures to cold, western drywood termites in regions of the wall that were actually chilled to $-20\,°C$ were killed. Some termites chilled to $<0\,°C$ but not to $-20\,°C$ survived. Therefore, reaching CT_{min} is critical for control. When introduced at 0.9 kg min^{-1} near the top of the wall, liquid N_2 provided $-20\,°C$ (CT_{min} for *I. minor*) within 7 min at the upper wall stud, 11 min at the midpoint, and in about 16 min at the lower wall stud (Fig. 7.1). Obviously, the time lag for cooling the bottom of the void was related to thermal dissipation. Being heavier than air, optimal rate of cooling was obtained when liquid N_2 was released high in the wall.

Rate of introduction of liquid N_2 into voids dramatically affected the time to reach Ct_{min}. Interestingly, no part of adjacent voids reached CT_{min}; the average time to CT_{min} in the base plate was 28 min when liquid N_2 was introduced at 0.5-0.9 kg min^{-1} into the top of the void, but was only 4 min when introduced at 2.3 kg min^{-1}. Again, rate of cooling was related to thermal

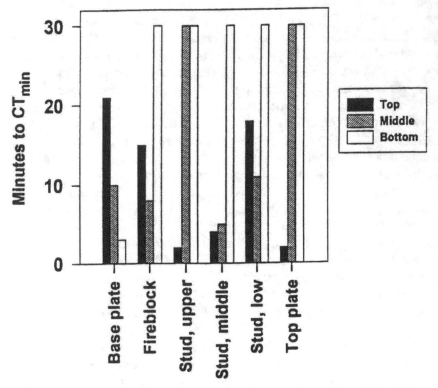

FIGURE 7.1 Effect of point introduction of liquid N_2 applied at 0.9 - 1.4 kg min[-1] to an uninsulated mock-up wall void on the time to reach the Critical Thermal Minimum temperature (CT_{min}) for drywood termites ($-20°C$) (Rust et al. 1995).

dissipation. Uniform chilling to $-20°C$ within 10 min was achieved when liquid N_2 was introduced at rates >1.4 kg min[-1]. These data point out that fireblocks or other impediments may provide a physical barrier for liquid N_2 and, at rates of application of <1.4 kg min[-1], prevent the nitrogen from cooling lower portions of a wall void. In those instances it may be best to introduce the N_2 below the obstruction.

In walls containing no insulation, points above the injection site failed to reach the Ct_{min} ($-20°$ C) within 30 min, the endpoint of our measurements (Fig. 7.2). Liquid N_2 introduced about mid-height provided CT_{min}'s at the middle and lower stud wall and in the fireblock and base plate, but not at the top. When introduced at the lower point of the wall void, only the base plate reached CT_{min}. In instances where the void was insulated with standard R-15 fiberglass sheet, the flow of liquid N_2 was deflected, thereby resulting in longer times needed for the top plate and lower portion of the void studs to reach CT_{min} (Fig. 7.2).

Liquid N_2 treatments of the wall void caused no apparent adverse effects on electrical and plumbing components placed in the wall. Neither soldered nor threaded pipe joints leaked during or after treatments. However, as a precautionary measure, water should be allowed to run or water pipes in areas to be exposed should probably be drained to prevent freezing and ice expansion which may damage older pipes or fittings. There was no change in the conductivity of the new electrical wire we exposed to liquid N_2 in our mock-up wall. Similar tests should be conducted with old wire.

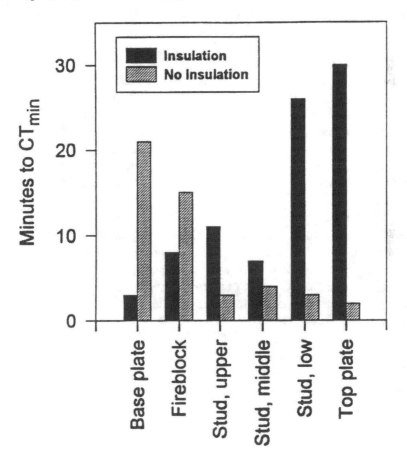

FIGURE 7.2 Effect of insulation on the time to reach the Critical Thermal Minimum temperature (CT_{min}) for drywood termites (–20°C). Liquid N_2 introduced at 0.9 - 1.4 kg min[-1] (Rust et al. 1995).

Safety should always be a concern whenever oxygen displacement is being used for control. The oxygen concentrations in the room and in the wall voids were monitored as liquid N_2 was introduced. The O_2 concentrations in the wall voids decreased to as low as 1.7%, but rose to 20.4% within at least 30 min after treatment. It is unlikely that anoxia contributed to the death of insects because Rust & Kennedy (1992) showed that continuous exposures of >96 h are needed to kill *I. minor* in atmospheres <0.1% O_2 at 25.5 °C. The O_2 concentration in the room where the tests were conducted routinely fell slightly below 18%, and fans were used as a precaution to ventilate oxygen into the room. When using liquid N_2, care should be taken to provide an adequate supply of O_2 whenever there is the possibility of nitrogen displacing vital oxygen for even a short period of time.

Lewis & Haverty (1996) tested the efficacy of liquid N_2 to freeze drywood termites in simulated infestations with similar results. Wall voids were treated with liquid N_2 at 382, 123, and 57 kg/m^3, and provided 100, 98.2, and 84.3% kill of all test insects in artificially infested boards, respectively. After exposure there were live termites in boards in about 30% of the wall voids treated, even though the thermocouples registered between -33 and -83 °C. This clearly suggested that thermocouple placement is crucial. Obviously, termites cannot survive exposure to that low temperature, so it is important to monitor the temperature until it drops to at least -20 °C directly where the termites are living.

The use of extreme low temperatures can, therefore, be an effective strategy to kill wood-destroying organisms. If a sufficient quantity of liquid N_2 is confined around infested objects, minimum lethal temperatures can be reached. However, precise monitoring of internal wood temperature is important and proper placement of thermocouples is crucial. Placement must reflect the environmental condition immediately surrounding the target pest. Because the effective distance of the cooling process is minimal and studs in adjacent wall voids do not reach CT_{min}, localized areas must be thoroughly bathed in liquid N_2 in order to sufficiently lower temperatures to a lethal level. Operational factors such as insulation mats and multiple points of N_2 introduction may increase the area effectively chilled.

Extreme High Temperature

Heat sterilization to disinfest grains has been practiced for nearly a century (Fields 1992). Sheppard (1984) recommended that food processing plants heat treat on warm days so that differences between indoor and outdoor temperatures are no greater then 28 °C. Factors that can seriously reduce the effectiveness of heat treatments in large structures such as warehouses, mills, and processing plants include large temperature gradients that develop from strong convection currents, inadequate air circulation, pockets of static air inside machinery, equipment, double walls, and the prevalence of unsealed openings in floors, walls, and roofs. Some version of these same conditions also exists in urban structures.

One measure of the maximum temperature tolerance of an insect is the critical thermal maximum (CT_{max}), the temperature at which irreversible torpor occurs. CT_{max}'s are determined by exposing the insect to slowly increasing temperature (usually about 1 °C min^{-1}) in water-saturated air, just to the point of knockdown, the point at which the insect succumbs. This provides an approximation of the maximum heat that an insect can tolerate for momentary exposures. Typically CT_{max} measurements have been made on desert species and only a few urban pests have been evaluated to determine their CT_{max}. Presumably it was thought unimportant because it was unlikely that urban insects would ever be exposed to lethal high temperatures. The CT_{max} of *I. minor* is 52°C (Rust et al. 1979). Appel et al. (1983) reported CT_{max}'s for the German cockroach, *Blattella germanica*, and the brownbanded cockroach, *Supella longipalpa*, being 48.7° and 51.4°C, respectively. Even brief exposures at these temperatures are lethal.

Museum Pests

Strang (1992) reported that the acute upper lethal temperature (ULT) for many museum pests ranged from 37-64°C depending on the length of exposure. The ULT differs from the CT_{max}'s in that the ULT invloves maximal tolerance to exposure to a high temperature for an extended period of time whereas the CT_{max} represents the maximal temperature an insect can tolerate if the temperature is gradually increased. Typically, the CT_{max} is higher than the ULT because of the brief (usually 1 min) exposure involved in determining the CT_{max}. Higher temperatures generally require shorter exposures. Most studies report the efficacy data from exposures to a specific high temperature and period, but detailed and systematic studies with longer exposures at lower temperatures have not been conducted with most insect pests. Rawle (1951) reported that 4-h exposures at 41°C killed all stages of webbing clothes moth, but 4-h exposures at 38°C failed to kill the egg. Interestingly, adult male moths exposed to 35°C for 2 days were sterilized whereas the females exposed to that temperature laid viable eggs, suggesting possible physiological different responses between sexes to elevated temperatures. Forbes & Ebeling (1987) reported on the effects of exposure to high temperatures in several urban insect pests exposed at ambient relative humidity (Table 7.2). Saturation eliminates evaporative cooling. Exposures at 51°C produced 100% kill of German cockroaches in 16 min, and increasing the temperature only 3°C reduced survival to ≤7 min.

Because it is unlikely that heat treatments will always be conducted in a water-saturated atmosphere, we conducted similar exposure tests at different relative humidities with three important structural pests (Table 7.3). In general, time to kill insects increases as relative humidity decreases. Gunn & Notely (1936) found that the ULT in moist air for the oriental cockroach, *Blatta orientalis*, and the American cockroach, *Periplaneta americana*, (24-h exposure) was 37-39°C, whereas some *B. germanica* survived for >24 h at 39°C. When exposed for 1 h in moist air, all three species were killed at 43°C; in dry air survivorship increased. Temperatures of at least 45°C were needed to kill all the cockroaches. Insects are capable of evaporative cooling for only short periods of time, owing to their relatively small size and high ratio of body surface area to volume (Edney 1977). Exposures at low humidity lead to death by desiccation, as shown in Fig. 7.3.

There are virtually no published reports concerning the use of microwaves to kill insects associated with artifacts or museum items. Exposure to microwaves kills eggs, larvae, and adult webbing clothes moths with minimal effects on color and tensile strength of wool (Reagan et al. 1980). Theoretically, insects exposed to microwaves may be heated to lethal temper-

TABLE 7.2 Time for 100% kill of four household pests at four temperatures in ambient relative humidity (Forbes & Ebeling 1987)

	LT_{100} (minutes)			
Temp. (°C)	Argentine ant (worker)	German cockroach (adult males)	Confused flour beetle (adult)	Western drywood termite (nymph)
46	8.0	58	123	265
49	4.0	27	16	33
51	2.5	16	9	10
54	1.0	7	4	6

TABLE 7.3 Effect of humidity on the time (minutes) for 100% kill of three selected structural insects exposed to high temperatures (Rust, Reierson & Roelofs unpublished data)

Pest and RH (%)	LT_{100} at						
	46.1 °C	48.9 °C	51.7 °C	54.4 °C	57.2 °C	60.0 °C	62.8 °C
German cockroach (adult male)							
100	-	30	30	15	15	-	-
49	60	60	60	30	15	15	-
11	60	60	60	45	30	30	15
Western drywood termite (worker)							
100	210	45	15	-	-	-	-
49	120	30	15	-	-	-	-
11	90	60	15	-	-	-	-
Carpenter ant *Camponotus* spp. (worker)							
100	180	15	15	-	-	-	-
49	120	45	30	-	-	-	-
11	120	30	30	-	-	-	-

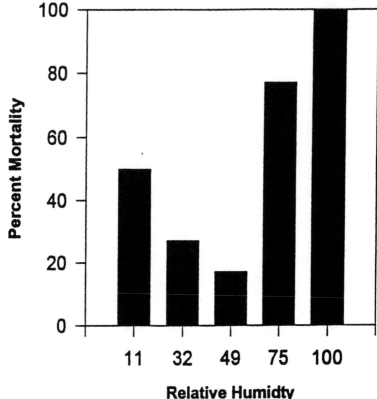

FIGURE 7.3 Percentage mortality of drywood termite workers exposed to various relative humidities at 43 °C for 2 h.

atures as their hemolymph and liquid in their tissues absorb the waves. Effective rates of application, methodology to direct waves to the target, and ways to minimize damage have yet to be determined. Lewis & Haverty (1996) found variable results with microwaves against drywood termites citing non-uniform heating as a possible problem. They reported some minor warping and burning of test boards. The effects of microwaves on any object depend on the amount of moisture the object contains. Therefore, the concept of disinfestation of very dry, non-metallic objects with microwaves needs additional investigation.

Household Insect Pests

Zeichner et al. (1996) reported that heat treatment of food service facilities provided outstanding control of German cockroaches. Prior to the heat treatment, one facility had received between 7 and 11 insecticide applications annually for three years. After heating, only four treatments were applied and the need for treatment, as indicated by a trapping index, declined below the treatment threshold (Fig. 7.4). While the costs for heating a 710 m² structure is somewhat expensive (ca. US $4,100), Zeichner et al. (1996) felt it was a good value considering the chronic history of insecticide resistance and the overall reduction in treatments and manpower.

In addition to direct effects of high temperatures on insects, Ebeling (1990) identified a phenomenon he labeled "heat synergism" whereby a combination of heat and insecticide had more of an effect than either heat or insecticide alone. He found that adult confused flour beetles died 6 days after being exposed to boric acid powder for 2 h at 46.1°C, whereas at 21.1°C only 5% died after being exposed to boric acid alone, and none died after being exposed to 46.1°C alone. The results were not as dramatic when adult male German cockroaches were exposed to heat + boric acid. Synergism of heat + insecticide suggests that (1) increased mobility of insects from high heat increases pick up of insecticidal boric acid, (2) heat changes the structure of the lipid wax layer and allows increased penetration of insecticide, or (3) heat increases the physiological activity of the insecticide after penetration. Ebeling (1995) also found that the desiccating properties of inorganic powders such as silica aerogels were increased by temperature. At 43.3°C, the time for 100% kill of German cockroaches with Dri-Die 67 and activated pyrethrum in silica aerogel (PT-230 Tri Die, Whitmire Research Laboratories, St. Louis, MO) decreased by 58 and 146 min, respectively.

Wood-Destroying Insects

Drywood termites represent one of the most important groups of wood-destroying insect pests in the world. They account for hundreds of millions of dollars spent annually for their control and for repair of damage they cause. Although distributed throughout geographic areas of seasonal hot conditions, xeric-adapted species such as *I. minor* can deal with dry conditions better than high heat. They can live for weeks without free water but die within 30 min if exposed to 49°C (Ebeling & Forbes 1988). Structural wood in buildings in semi-desert and southern temperate regions sometimes naturally heats to temperatures exceeding 49°C, a threshold maximum temperature above which termites and other wood-destroying insects cannot live for long. This is especially true for sheathing, attic rafters, and studs in exposed walls during the summer and fall when heat is most intense. Daily monitored temperatures for a year in structural wood in a building in Riverside, California revealed

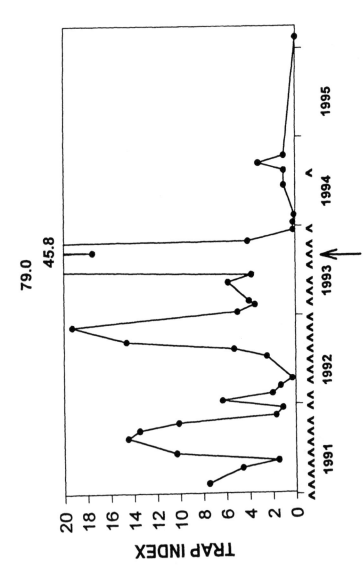

FIGURE 7.4 Efficacy of a mid-September heat treatment in a food preparation facility to control German cockroaches. The target temperature in the structures was 46°C for 45 min. Small arrows on the horizontal margin indicate insecticide treatments; large arrow indicates heat treatment (Modified from Zeichner et al. 1996)

instances where temperatures exceeded 54°C, albeit usually for less than an hour or two, on some hot summer days (Rust & Reierson, unpublished data). Termites and other pests in wood must avoid those excessive temperatures in order to survive, and they may accomplish this either by remaining in portions of wood protected from high temperatures or by temporarily moving away from high heat into cooler parts of the wood as the temperature increases to lethal levels. Termites exposed to temperatures above 52°C for even a brief time will die. We have reported this phenomenon of movement in response to heat for the drywood termite, *I. fruticavus*, in southern California (Rust et al. 1979). The lethal susceptibility of insects to heat suggests that high heat (i.e. temperatures above their CT_{max}), coupled with an inability to escape, may provide an effective strategy for controlling wood-destroying pests.

Operationally, structural temperatures can be raised to levels that will kill termites by heating the air with special propane heaters. In some cases the entire structure is heated and in others only localized areas are heated. The center of a sill or bottom plate resting on concrete can be heated to 49°C in about 6 h. Typically, a sill plate is the last place in the home to heat up to lethal temperature. Potential problems encountered when heating structures include melting of plastics, especially if the plastic is on or near metal such as clamps or sprinkler head riser tubes adjacent to outside walls. This can be minimized by using thermal blankets and moving plastic away from metal (Ogle 1990). When heating food handling facilities, Zeichner et al. (1996) found that ceramic tiles not firmly cemented to concrete floors were lifted or cracked. Older models of refrigerators/freezers with door panels that are pressure fitted together as opposed to welded were removed because they may become distorted due to the expansion of foam in the door.

Snyder (1923) reported that seasoned woods need to be kiln dried for at least 30 min at 82°C to kill larvae of powderpost beetles. Temperatures below 54.4°C will not kill larval powderpost beetles, but larvae in infested boards are killed if the wood is heated with steam to 54.4°C and held at that temperature for 1.5 h (Snyder & St. George 1924). Ebeling et al. (1989) successfully controlled an unidentified species of powderpost or anobiid beetle in three different structures by heating the wood to 49°C. In subsequent studies, infested blocks heated to 54°C for 30 min killed the immatures and prevented production of frass. Studies we have made with *Lyctus planicollis* indicate that naked larvae died from a 60-min exposure to 51.7°C. Exposures to 51°C, 55-56°C, and 51-52°C for only 5 min killed larval *L. brunneus, L. africanus*, and the deathwatch beetle, *Anobium punctatum*, respectively (Cymorek 1972). Latent lethal effects may result from exposure to high heat, but that phenomenon has not been well-documented. Toskina (1978) reported that exposure at 30°C induces thermal torpor in adult female deathwatch beetles, and after several days, females no longer lay eggs. Exposures at 34°C were lethal to eggs. Additional temperature exposure studies are needed to determine the precise minimum lethal temperatures (MLTs) for these and other important species of wood-destroying beetles and to determine the relevant long-term effects that minimum high heat may have on insect pests.

Surrounding humidity and internal moisture content affects the heating characteristics of wood. In tests where we heated Douglas fir cubes (562 cm³) in an oven under a variety of conditions, we found that blocks with the highest percentage moisture and those heated while exposed at the highest relative humidity gained heat the slowest (Table 7.4). While moist blocks at 100% humidity gained 19°C, blocks heated at 0% humidity gained 22°C within 60 min. For any given heating process, it will take longer to attain MLT as wood dimensions increase, for wood containing a high percentage of water, and for wood being heated under high humidity conditions.

The use of heat to control wood-destroying insects in structural lumber looks extremely promising. The inability of these insects to survive even short exposures to temperatures above 52 °C makes the heating strategy practical under many conditions.

TABLE 7.4 Effect of moisture on the heating characteristics of Douglas fir wood blocks (8.3 cm cubes) conditioned at each RH for 6 months and then heated in a preheated oven at 49.8°C (Rust, Reierson, & Roelofs unpublished data).

RH (%)	% Wood Moisture (± SD)	Average temp. (°C) gain in the center at		
		10 min	30 min	60 min
0	2.6 ± 0.3	7.1	16.6	22.2
50	6.9 ± 0.68	3.6	13.2	18.4
100	19.2 ± 0.66	1.2	7.0	12.6

Future Directions

The use of extreme temperatures to control insects in and around structures is very promising. The fact that structural insect pests die at subzero and high temperatures is proven. It is somewhat surprising, however, that many kinds of structural insects are unable to withstand even marginally extreme temperatures for relatively brief periods of time. Delivery techniques are being developed to optimize control with freezing and heat. In fact, both freezing and heating strategies have been successfully commercialized to varying levels of success. Additional promotion of the use of the heat processes to control structural pests is needed. More refined studies of the effects of temperature and humidity on structural pests should be conducted. In addition to acute lethal effects, studies of chronic, delayed, or latent effects of sublethal exposures to extreme temperatures may provide interesting results and add to a more complete understanding of the effectiveness of these techniques. It is essential to monitor treated objects to insure that adequate lethal temperatures have been achieved. Additional technological advances may simplify the monitoring procedure. An obvious advantage of treating with cold or heat in sensitive areas such as food handling facilities and museums is the lack of pesticide residue. Although freezing and high heat technologies presently require specialized bulky equipment and awkward apparatus, it is likely that continuing pressure to develop safe and effective control strategies involving less application of pesticides will lead to refinement of equipment and procedures that will make freezing and heating even more relevant. Highly portable thermal delivery systems (high-velocity hot-air blowers for example) or ultra-cold blast freezer units are being developed and will hopefully help make this technology even more widely used and accepted. Portable hand-held equipment to control cockroaches and other household pests would be extremely attractive to consumers and pest control technicians. Use of extreme temperature is likely to become an increasingly important component of a variety of urban integrated pest management programs in the future.

References

Appel, A. G., D. A. Reierson & M. K. Rust. 1983. Comparative water relations and temperature sensitivity of cockroaches. Comp. Biochem. Physiol. 74A: 357-61.

Asahina, E. 1966. Freezing and frost resistance in insects *In* H.T. Meryman ed. *Cryobiology*, pp 451-86. Academic Press. London.

Back, E. A. 1935. Clothes moths and their control. U. S. Department of Agriculture Farmer's Bulletin No. 1353. 29 pp.

Back, E. A. & R. T. Cotton. 1927. Effect of cold storage upon clothes moths. Refrigerating Engineering 13: 365-6.

Baust, J. G. & R. R. Rojas. 1985. Review- insect cold hardiness: facts and fancy. J. Insect Physiol. 31: 755-9.

Brokerhof, A. W., R. Morton & H. J. Banks. 1993. Time-mortality relationships for different species and developmental stages of clothes moths (Lepidoptera: Tineidae) exposed to cold. J. Stored Prod. Res. 29: 277-82.

Chen, C.-P., D. L. Denlinger & R. E. Lee, Jr. 1987. Cold-shock injury and rapid cold hardening in the flesh fly *Sarcophaga crassipalpis*. Physiol. Zool. 60: 297-304.

Chen, C.-P., R. E. Lee, Jr. & D. L. Denlinger. 1991. Cold shock and heat shock: a comparison of the protection generated by brief pretreatment at less severe temperatures. Physiol. Entomol. 16: 19-26.

Cymorek, S. 1972. Uber notewendige Temperatur-Zeit-Verhaltnisse zum Abtoten von *Lyctus*-Larven bezogen auf die Heissverleimung von Turen. Mitteilungen Deustchen Gesellschaft fur Holzforschung 57: 50-7.

Darlin, D. 1994. A chilling tale. Forbes Mag. July: 116-7.

Denlinger, D. L., K. H. Joplin, C.-P. Chen & R. E. Lee, Jr. 1991. Cold shock and heat shock, *In* R. E. Lee, Jr. & D. L. Denlinger eds. *Insects at Low Temperatures,* pp 131-148.. Chapman and Hall, New York and London.

Ebeling, W. 1990. Heat and boric acid: an example of synergism. Pest Cont. Tech. 18: 44, 46.

Ebeling, W. 1995. Inorganic insecticides and dust, *In* M. K. Rust, J. M. Owens & D. A. Reierson eds. *Understanding and Controlling the German Cockroach,* pp 193-230. Oxford Univ. Press, New York. 430 pp.

Ebeling, W. & C. F. Forbes. 1988. Use of high temperature for the elimination of drywood termites, p 453. Proc. XVIII International Congress of Entomology, Vancouver.

Ebeling, W., C. F. Forbes & S. Ebeling. 1989. Heat treatment for powderpost beetles. The IPM Practitioner 11: 1-4.

Edney, E. B. 1977. *Water Balance in Land Arthropods.* Springer-Verlag, Berlin. 282 pp.

Fields, P. G. 1992. The control of stored-product insects and mites with extreme temperatures. J. Stored Prod. Res. 28: 89-118.

Florian, M.-L. E. 1986. The freezing process - effects on insects and artifact materials. Leather Conservation News 3: 1-13.

Florian, M.-L. E. 1990. Freezing for museum insect pest eradication. Collection Forum 6: 1-7.

Florian, M.-L. E. 1992. Saga of the saggy bag. Leather Conservation News 8: 1-11.

Forbes, C. F. & W. Ebeling. 1986. Update: liquid nitrogen controls drywood termites. The IPM Practitioner 8: 1-4.

Forbes, C. F. & W. Ebeling. 1987. Update: use of heat for elimination of structural pests. The IPM Practitioner 9: 1-6.

Gilberg, M. & A. Brokerhof. 1991. The control of insect pests in museum collections: the effects of low temperature on *Stegobium paniceum* (Linnaeus), the drugstore beetle. J. Am. Inst. Conservation 30: 197-201.

Gunn, D. L. & F. B. Notley. 1936. The temperature and humidity relations of the cockroach. IV. Thermal death-point. J. Exp. Biol. 13: 28-34.

Kite, M. 1992. Freezing test of leather repair adhesives. Leather Conservation News 7(2): 1-2.

Kronkright, D. 1992. Experiences with freezing as a method of insect eradication in museum collections with special emphasis on wooden artifacts. Proceedings Wooden Artifacts Group, American Institute for Conservation.

Lee, R. E., Jr. 1991. Principles of insect low temperature tolerances, *In* R. E. Lee, Jr. & D. L. Denlinger, eds. *Insects at Low Temperatures,* pp 17-46. Chapman and Hall, New York and London.

Lewis, V. R. & M. I. Haverty. 1996. Evaluation of six techniques for control of the western drywood termite (Isoptera: Kalotermitidae) in structures. J. Econ. Entomol. 89: 922-34.

Miller, L. K. 1978. Freeze tolerance in relation to cooling rate in an adult insect. Cyrobiology 15: 345-9.

Mullen, M. A. & R. T. Arbogast. 1984. Low temperatures to control stored-product insects, *In* F. J. Baur, ed. *Insect management for Food Storage and Processing,* pp 255-63. American Assoc. Cereal Chemist, St. Paul, Minnesota. 384 pp.

Nesheim, K. 1984. The Yale non-toxic method of eradicating book-eating insects by deep-freezing. Restaurator 6: 147-64.

Ogle, J. 1990. Using new heat technology for termite treatment. Voice of Pest Control Operators of Calif. 27, 37.

Rawle, S. G. 1951. The effects of high temperature on the common clothes moth, *Tineola bisselliella* (Humm.). Bull. Entomol. Res. 42: 29-40.

Reagan, B. M., J.-H. Chiao-Cheng & N. J. Street. 1980. Effects of microwave radiation on the webbing clothes moth, *Tineola bisselliella* (Humm.) and textiles. J. Food Protection 43: 658-63.

Rust, M. K. & J. M. Kennedy. 1992. The feasibility of using modified controlled atmospheres to control insect pests in museums. Getty Conservation Institute, Marina del Rey, CA. 126 pp.

Rust, M. K., E. O. Paine & D. A. Reierson. 1995. Laboratory evaluation of low temperature for controlling drywood termites. California Department of Consumer Affairs Report. 61 pp.

Rust, M. K., D. A. Reierson & R. H. Scheffrahn. 1979. Comparative habits, host utilization and xeric adaptations of southwestern drywood termites, *Incisitermes fruticavus* Rust and *Incisitermes minor* (Hagen) (Isoptera: Kalotermitidae). Sociobiology 4: 239-55.

Salt, R. W. 1961. Principles of insect cold-hardiness. Ann. Rev. Entomol. 6: 55-74.

Shibasaki, M., H. Ikeda, S. Isoyama, N. Imoto, K. Takeda, E. Noguchi & H. Takita. 1996. Treatment of whole houses with liquid nitrogen for control of dust mites. J. Med. Entomol. 33: 906-910.

Sheppard, K. O. 1984. Heat sterilization (superheating) as a control for stored-grain pests in a food plant, *In* F. J. Baur, ed. *Insect Management for Food Storage and Processing*, pp. 194-202. American Assoc. Cereal Chemist, St. Paul, Minnesota.

Smith, R. D. 1984. Background, use, and benefits of blast freezers in the prevention and extermination of insects. Biodeterioration 6: 374-9.

Snyder, T. E. 1923. High temperatures as a remedy for *Lyctus* powder-post beetles. J. Forestry 21: 810-4.

Snyder T. E. & R. A. St. George. 1924. Determination of temperatures fatal to the powder-post beetle, *Lyctus planicollis* LeConte, by steaming infested ash and oak lumber in a kiln. J. Agric. Res. 28: 1033-9.

Strang, T. J. K. 1992. A review of published temperatures for the control of pest insects in museums. Collection Forum 8: 41-67.

Strang, T. J. K. 1995. The effect of thermal methods of pest control on museum collections, pp. 199-212. 3rd International Conference on Biodeterioration of Cultural Property, 4-7 July 1995, Bangkok, Thailand.

Tanimura, H. & S. Yamaguchi. 1995. The freezing method for eradication of museum pest insects - the safe method for both the human body and artifacts. Experiments on Japanese artifacts and the present state of freezing method in western countries, *3rd International Conference on Biodeterioration of Cultural Property*, pp 387-98. 4-7 July 1995, Bangkok, Thailand.

Toskina, I. N. 1978. Wood pests in articles and structures and pest control in museums, *ICOM Committee for Conservation 5th Triennial Meeting*, pp 1-10. Zabreb.

Williams, S. L., H. H. Genoways & D. A. Schlitler. 1985. Control of insect pests in mammal collections. Acta Zool. Fennica 170: 71-3.

Zeichner, B. C., A. L. Hoch & D. F. Wood, Jr. 1996. The use of heat for control of chronic German cockroach infestations in food service facilities - fresh start,*The 2nd International Conference on Insect Pests in the Urban Environment*, pp 507-513. 7-10 July 1996, Edinburgh, Scotland.

8

Temperature Treatments for Quarantine Security: New Approaches for Fresh Commodities

Robert L. Mangan and Guy J. Hallman

The accidental spread of organisms through human migration or commerce is generally recognized as "unnatural" or undesirable for ecological as well as economic reasons. Phytosanitary quarantines are imposed on a wide variety of plants and plant products as a means to deter introductions of pathogens and pests. There is nearly universal agreement among commercial, ecological, and scientific interests that quarantines are desirable to limit increased costs of agricultural production due to new pests and to reduce risks of environmental degradation by undesirable invasions of exotic organisms.

In contrast with the desirability of protecting agricultural and natural ecosystems from pests and exotics, quarantine implementation is associated with undesirable effects including restriction of commodity availability, increased costs, and decreased commodity quality. Fresh fruits and vegetables are components of healthy lifestyles and are promoted as such. The fumigant ethylene dibromide was withdrawn from use in the mid-1980's because of concerns with carcinogenicity, and methyl bromide has been associated with stratospheric ozone depletion and may be lost as a quarantine treatment in a few years. Public concerns about cancer or environmental degradation conflict with the healthy benefits associated with fresh produce. Food safety as well as mere public perceptions of food safety are among the highest concerns in development of quarantine treatments for fresh fruits and vegetables. For example, although gamma irradiation has been shown to be a viable quarantine treatment with no food safety problems posed by the irradiated produce itself, opposition to food irradiation from a segment of the public has caused commercial and governmental sectors worldwide to proceed with great caution and delay in adopting irradiation quarantine treatments.

Quarantine treatments of fresh fruits and vegetables involve killing, removing, or preventing reproduction by an undesirable living organism on living host material. For example, fruit fly quarantine treatments usually involve killing eggs or larvae that are developing inside the living host. This system requires that the insect receive a lethal treatment inducing very high (usually sufficient to kill 99.9968% of pests) mortality while the plant tissue is minimally affected (Shannon 1994). In addition, the treatment must have reasonable cost and minimally encumber the marketing system. Solving the complex problem of attaining a very high level of pest mortality while preserving the living commodity and the economic, engineering, and logistics considerations in treating large volumes of produce during a restricted and busy harvest season is the challenge of quarantine treatment research on fresh commodities.

The process of establishing new quarantine treatments is composed of three components. First, there is an evaluation of the risk of pest introduction. If the pest already occurs in the importing region or it is determined that the pest cannot survive in the importing region risk may be negligible and treatment not required. The range of risk for pests requiring some action may range from inspection (for "hitch-hiking" pests not normally found on the commodity) to disinfestation treatment. Should a postharvest treatment be required, the second step is development of an effective treatment. Normally the development is divided into three parts, establishment of the information base to determine the feasibility of a proposed treatment (host tolerance, cost, method of implementation) and determination of the required treatment dose. A treatment is determined to be effective when it is confirmed that it actually achieves the required mortality rates. The third component is development of an operational system to manage the treatment as part of the quarantine system. This involves establishing a work plan that sets quality control/assurance standards for operating the system and defines the actual operation of facilities.

Review of Temperature Quarantine Treatments

Extreme temperatures were the first quarantine treatments. Ironically, cold storage was initially implicated in the spread of insect pests before it was used as a quarantine treatment. Fuller (1906) in South Africa, noted that fruit fly larvae survived 124 days in peaches stored at about 4.5°C and considered this evidence that international shipment of cold-stored fruit may be responsible for the spread of pests. Studies in Australia determined that fruit flies survived temperatures above 3.3°C, but that temperatures between 0.5-1.7°C for three weeks killed all fruit fly eggs and larvae in various fruits (Anonymous 1907). Subsequent research on the effects of cold temperatures on fruit fly larvae were dedicated to ensuring that the cold storage method used did not allow for live insects to be moved in cold-stored fruit (Gould 1994). Later, cold was investigated as a quarantine treatment in its own right. As early as 1914 it was proposed that citrus be heated to disinfest it of Mexican fruit fly, *Anastrepha ludens* (Baker 1952). Early research on the susceptibility of larvae of that fly to heat was a prominent impetus to the use of the vapor heat treatment against Mediterranean fruit fly, *Ceratitis capitata,* in Florida. Exposure of citrus to saturated 43.3°C air for 14-16 h or 1.1°C for 12 days were used to disinfest many tons of Florida citrus during the first Mediterranean fruit fly outbreak in that state that lasted for 19 months beginning in 1929 (Baker 1952, Richardson 1952).

Soon thereafter cold storage and vapor heat were being used as quarantine treatments in numerous settings (Gould 1994, Hallman & Armstrong 1994). However, by the early 1950's fumigants had proven to be easy, effective, and cheap means of disinfesting fresh commodities, and vapor heat was no longer used as a quarantine treatment. Use of cold storage as a quarantine treatment declined dramatically, also. This situation persisted until the late 1970's when the Japanese began to use vapor heat again because of fumigant damage to some fresh commodities (Hallman & Armstrong 1994). In the mid 1980's the fumigant ethylene dibromide was banned as being carcinogenic (Ruckelshaus 1984). In response, immersion of mangoes in 46.1°C water for up to 90 min was developed as a quarantine treatment to replace ethylene dibromide fumigation (Sharp 1994), and interest was renewed in the vapor heat treatment. Improvements in design of vapor heat treatment equipment, such as higher air speed forced through the fruit load, better temperature control which allowed for increased treatment temperatures, and increased reliance on dose-mortality statistics to more closely determine doses needed for treatment efficacy enabled treatment times to be drastically reduced from the 14-16 h commonly used to about 5 h (Hallman & Armstrong 1994). The relatively newer treatment known variously as "high temperature forced air", "forced heated air", "forced hot air", or simply "hot air" is very similar to the old vapor heat treatment except air is forced through the commodity load (Gaffney & Armstrong 1990). Although some differences in definitions occur the two treatments overlap into a category which can be called heated air quarantine treatments (Hallman & Armstrong 1994). The name used is not as important as clearly defining what was done during the treatment regarding ranges of temperature, humidity, air speed, time, and any other variables which may affect efficacy and fruit quality.

The latest occurrence to spur the development of temperature-based and other alternative quarantine treatments is the impending loss of the last fumigant used on fresh commodities, methyl bromide.

Research effort in quarantine treatments has increased dramatically since 1980. We have summarized the numbers of publications that include the terms "quarantine" and "fruit or vegetables or flowers" from the Agricola data base (compiled by the U.S. National Agriculture Library) from 1970 to the present in Fig. 8.1. We interpret the increased publication rate to be a result of several developments: (1) restrictions placed on fumigants used for quarantine security, (2) removal of international trade restrictions, (3) improvements in technology for transporting fresh plant commodities, (4) the increasing popularity of exotic fresh commodities, and (5) a trend toward smaller publishable units. All but the last point deserve further discussion.

Ethylene dibromide was banned in the U.S. in 1984 as a health risk (Ruckelshaus 1984), and phaseout of methyl bromide is being scheduled worldwide because it is reported to be a significant stratospheric ozone depleter (Anonymous 1993). When ethylene dibromide was withdrawn, methyl bromide replaced it as a quarantine treatment for many commodities. At present there is no fumigant to replace methyl bromide for quarantine treatment of fresh commodities. The Montreal Protocol has recently exempted the use of methyl bromide as a quarantine treatment. However, its continued production or importation in the United States (U.S.) past January 1, 2001, would still violate the U. S. Clean Air Act, a law that would have to be amended for the continued legal production or importation of methyl bromide to occur in the U. S. after that date. Restrictions on use of the fumigant in the U. S. may extend to commodities

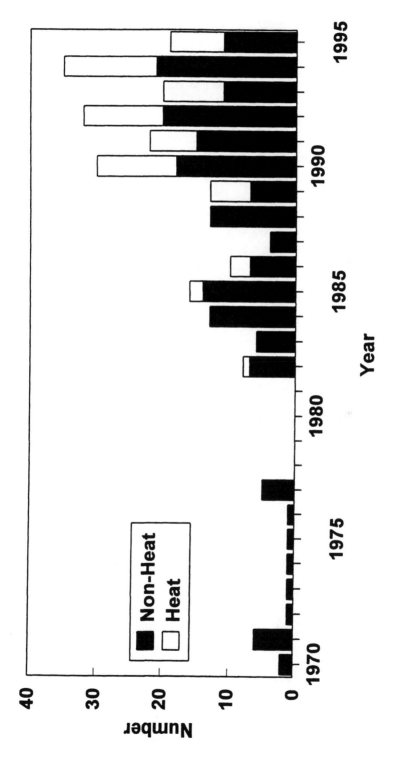

FIGURE 8.1 Numbers of quarantine-related publications in the Agricola data base per year.

exported from other countries to the U. S. as well. Regardless of whether or not methyl bromide fumigation survives as a quarantine treatment, public and retailer opposition to the use of any fumigant in fresh commodities for human consumption may eventually halt its continued use (Kelly 1996).

A trend in relaxation of tariffs as trade restrictions has made international marketing of perishable commodities a more profitable business. However, this softening of trade barriers has resulted in greater scrutiny by agricultural pest regulatory agencies. Thus, the need for treatments to solve new quarantine problems. Additionally the recent North American Free Trade Agreement (NAFTA) and General Agreement on Trade and Tariff (GATT) specify that phytosanitary restrictions on trade are to be based on scientific considerations and both agreements encourage or require research to overcome these barriers to market access. Article 762 (1) of NAFTA states:

"Each party shall, upon the request of another Party, facilitate the provision of technical advice, information and assistance, on mutually agreed terms and conditions, to enhance that Party's sanitary and phytosanitary measures and related activities including research, processing technologies, infrastructure and the establishment of national regulatory bodies."

The text of the GATT (now part of the World Trade Organization) agreement is more specific. Article IIA sec 1A #4, para. 29-30 states:

"29. Members agree to facilitate the provision of technical assistance to other Members, especially developing country Members, either bilaterally or through appropriate international organizations. Such assistance may be, inter alia, in the areas of processing technologies, research and infrastructure...."

And

"30. Where substantial investments are required in order for an exporting developing country Member to fulfil the sanitary or phytosanitary requirements of an importing Member, the latter shall consider providing such technical assistance as will permit the developing country Member to maintain and expand its market access opportunities for the product involved."

Publication of research on quarantine technology, therefore, benefits domestic production and marketing within a country by providing means to protect against introduction of exotic pests in commodities that the region wishes to import (such as Mexican mangoes entering southern Califorma and Texas) or allowing minimal market interruption if exotic pests have outbreaks in these regions. This research also fulfills conditions in the trade agreements, especially in relations with less developed countries.

Improvements in the technology of food transport, such as controlled atmospheres en route, have resulted in the ability to ship products of consistently good quality, which reduces losses while ensuring a steady market. Also, commodities which could not be shipped economically before because of poor shipping ability are being shipped more frequently due to improved postharvest technology (Buchanan 1994).

Consumers worldwide are eating a wider variety of fresh produce than before, some which are hosts of quarantined pests. For example, the increasing popularity of carambolas has created the necessity for quarantine treatments against pests of that fruit (Gould & Sharp 1990, Armstrong et al. 1995b). Countries such as Japan, Singapore, and Hong Kong have also greatly increased imports of fresh commodities over the last 10-15 years. According to Minnis (1994) Singapore, Hong Kong, Taiwan, and Japan have increased imports of the majority of fresh fruits since the mid 1980's, despite the presence of duties. Mango imports into Japan, Singapore, and Taiwan, for example, have more than doubled from the mid 1980's to 1992. Buchanan (1994) and Minnis (1994) attribute these increases to raises in disposable income and the cultural values assigned to fresh fruit. Proctor & Cropley (1994) showed a similar increase in fresh fruit imports into Europe; mango and papaya imports increased by 71 and 88%, respectively, during the 1987-1992 period.

Most of the new quarantine treatments are temperature based. Besides the examples already mentioned, heated air treatments are used against pests of lychees, papayas, mangoes, green peppers, and other fruits imported by Japan, and heated air-treated papayas are shipped from Hawaii to the continental United States (Hallman & Armstrong 1994). Grapefruits stored at cold temperatures for various days are shipped from Florida to Japan (Gould 1994).

Concerns with Temperature Treatment Research

A failure of a quarantine treatment can result in the establishment of a destructive exotic pest. The least it can do is cause disruption of shipment and economic loss to those involved in growing and marketing a product. Quarantine researchers do not have the luxury of being able to make somewhat impulsive conclusions about the success of a treatment, something which can be done without severe consequences in some other areas of pest managment where virtually total control is not needed and the failure of any single process may not result in significant loss. Quarantine treatments demand a very high level of security, and when they are used on a commercial scale treatments that do not achieve the level of quarantine security needed may be discovered to fail. Because of the extremely low level of infestation present in some commodities, that failure may not become manifest for some time. Consequently, there are many considerations when researching quarantine treatments. We discuss some that have caused problems or have been of concern. There may be many others. Researchers are encouraged to imagine their treatments being applied on a commercial scale and anticipate possible problems.

Assessing Mortality

Even something so seemingly simple as assessing mortality can cause problems for quarantine treatment research. Two different approaches have been used for infesting fruit for quarantine treatment research: (1) fruits are exposed to insect oviposition and the insects are allowed to develop to the appropriate stage; after treatment, fruits are stored to allow any survivors to emerge (Gould & Sharp 1990, Mangan & Ingle 1992), or (2) insects are reared on a diet and placed in fruits at the desired stage; after treatment, insects are removed from the fruits and mortality is assessed (Armstrong et al. 1989, Mangan & Ingle 1994). The first approach more closely approximates the natural condition, whereas the second approach is advantageous in that numbers of insects tested are known, stage and infestation rate can be precisely controlled, and infestability of the commodity is not a problem. Studies comparing the two treatments have not been done, however. Artificial infestation techniques should be compared with natural fruit infestation for differences in insect mortality during treatment.

There has been some variability in what criterium comprises mortality in a temperature treatment. Many researchers considered any larvae that emerged from infested fruits after treatment as survivors (Hallman et al. 1990). Others counted only puparia (Heather et al. 1996, Yokoyama & Miller 1996). Still others counted only normal appearing puparia (Gould & Sharp 1990, Sharp & Hallman 1992). Some researchers did not count puparia as survivors if no adult emerged. Failure to move several days after treatment even following prodding with pins and bristles was used by Whiting et al. (1991, 1995) to determine mortality after combination heat-modified atmosphere treatments. Still other researchers gave no clear definition of mortality. We feel that any insect, regardless of life stage, that emerges from the fruit or is alive when insects are removed from the fruit, in the case of artificial infestation, should be counted as survivors. This is because quarantine regulators who examine fruit will count any live insects as survivors; they will not wait for pupation let alone adult emergence. Furthermore, an insect which survives a severe temperature quarantine treatment may complete development and reproduce. Thomas & Mangan (1995) found that normal adults emerged from some abnormal tephritid puparia formed by heated last instars. Taschenberg et al. (1974) found 27% Caribbean fruit fly emergence from larviform puparia that were immersed as third instars in ambient water for 30 h. Hallman (1990a) found no relationship between percentage mortality of Caribbean fruit fly larvae subjected to heat in fruits and adult development from surviving larvae. Adults which survived as third instars a heat treatment that killed up to 99% of the larvae still reproduced about as well as adults from unheated larvae.

Temperature Control

The temperatures used for quarantine treatments (0-3 and 43-49°C) are on the border between temperatures which will not kill all insects in the amount of time used for treatments (>3 and <43°C) and those that might harm fresh commodities (<0 and >49°C). In these border areas changes of less than one degree can result in very different insect control levels and commodity responses. Therefore, precise temperature measurement and control are of utmost importance in quarantine temperature treatments. Recording equipment must be suitable for the task and calibrated with certified standards at the temperatures to be used in the research. Enough temperature measuring devices, such as thermocouples, must be present in the treated fruit load to be able to determine the range of temperatures experienced. Especially in heated air treatments, there may be areas which are significantly cooler than the treatment temperature because air will take the path of least resistance.

Humidity Level

The humidity level in heated air treatments has two important effects: (1) it influences commodity heating rate and, therefore, insect mortality rate, and (2) it affects commodity quality. Shellie & Mangan (1994) found faster heating rates of mangoes and grapefruits subjected to heated air near 100% RH compared with lower RH (Fig. 8.2). Humidity has long been

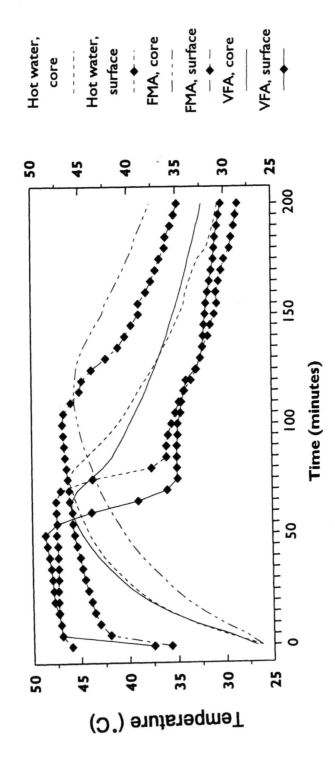

FIGURE 8.2 Temperature profiles of grapefruits subjected to three 48°C treatments until the center reached 46°C. Vapor heat was about 100% relative humidity (RH); forced moist-air was 60-90% RH (dewpoint was maintained at 2°C below dry bulb temperature) (after Shellie & Mangan 1994).

recognized as an important factor in fruit quality for heated air treatments of papayas. Papayas fared better when the air was not saturated during lengthy heat treatments compared with those heated under high relative humidity conditions. Jones (1939) noted that water condensed on papayas treated under high relative humidity and suggested that this may limit the absorption of oxygen by the fruits resulting in damage.

Genotype of Insect

Insects that have been reared in the laboratory for several generations are often different from their feral counterparts in many respects, such as length of life cycle, preovipositional period, fecundity, and flight ability, to name a few (Chambers 1977). Most quarantine treatment research is done with laboratory-reared colonies because of the need for large numbers of insects of the same stage. One obvious effect of genotype on cold efficacy is diapause. A colony which has genetically lost the ability to diapause will, of course, be more susceptible to cold in the diapausing stage than feral insects. Other than diapause little information exists comparing laboratory and feral insects for response to quarantine treatment. Sharp (1988) compares studies of hot water immersion of fruit fly-infested mangoes where wild and laboratory-reared flies were used. The results of probit analysis to estimate the $LT_{99.9968}$ show a small overlap in 95% fiducial limits for laboratory-reared West Indian fruit fly in Haiti (60-78 min) versus feral Mexican flies (74-98 min) in mangoes of similar sizes. On the other hand, $LT_{99.9968}$ estimates were comparable for laboratory-reared Mexican fruit flies in Texas and feral flies in Mexico. However, there were enough differences in these studies that could have influenced mortality, coupled with the insecurity in predicting a value as extreme as $LT_{99.9968}$, to make it impossible to use these data to answer a purely genetic question. In a more controlled series of experiments where oriental fruit fly third instars were removed from field-infested papayas (feral) or taken from a laboratory colony reared on semi-artificial diet and placed in the center of papayas subjected to heated air, no difference in mortality between the two fly types was evident (Hansen et al. 1990).

Because genetics ultimately controls the ability of organisms to express tolerance to any environmental stress and laboratory-reared insects are usually maintained at constant moderate temperatures, it is conceivable that laboratory-reared insects might differ in their response to heat and cold quarantine treatments compared with feral insects. Chapters 2 and 3 cite successful genetic selection studies for both cold and heat tolerance in *Drosophila* spp. This potential risk should be examined for pests of quarantine significance considering that at the moment little data exist.

Insect Population Density

Another difference between quarantine research and its commercial application that might cause concern is pest density. Quarantine treatment research is often done with many insects per fruit while the number of insects per fruit in commercial shipments may be very low. (In some cases, such as Mexican and oriental fruit flies, although the infestation rate may be extremely low, the number of insects found in an infested fruit may be high.) Hansen & Sharp (1997) found that in a system using 5-cm long pieces of 2.5-cm diameter metal conduit as a larvae-holding vehicle, Caribbean fruit fly third instar mortality increased significantly after more than about 50 larvae reared on diet were placed in the conduit and heated at 40°C for 67 min. However, quarantine treatment research has not been done at these high densities (or low temperatures) where mortality increased dramatically: over 50 insects per 24.5 cm^3 of host material. Hallman & Sharp (1990b) present data where the infestation density of Caribbean fruit fly third instars in carambolas immersed in 46-46.4°C varied from a mean of 2.8 to 34.0 per fruit. At the two lowest densities, mortality was 100% after 20 min, while at the two highest densities mortality was 97.8 and 99.9% after the same amount of time, which indicates that density up to 34 larvae per carambola did not cause increased mortality compared with lower densities. In any case, it is advisable to consider the possible effect of insect density when infesting fresh commodities for quarantine treatment research, especialy with insects which are evenly distributed, such as codling moth (Soderstrom et al. 1996b).

Temperature Regime in Research vs. Commercialization

The temperature regime to which insects are exposed before treatment can affect their response to that treatment. In most quarantine treatment research, the insects have been reared for many generations under controlled conditions which are in many ways different from those they face in their natural setting (Hallman 1996). Rearing temperatures are usually constant and moderate while temperatures of fruits in the field may fluctuate between quite cool and very warm. Hallman (1994b) found that Caribbean fruit flies reared at a constant 30°C were more tolerant of hot water immersion than those reared at constant lower temperatures. The effects of heat-shock proteins are well known and have been heavily studied in recent years (see Chapter 2). Holding last instar Queensland fruit flies, *Bactrocera tryoni*, at 35°C for 8 h prior to immersion in 46°C water for 4.5 min reduced mortality by more than two-thirds compared with larvae maintained at 25°C before immersion (Beckett & Evans 1997). Lester & Greenwood (1997) found that mild heat pre-treatments increased tolerance of lightbrown apple moth larvae to subsequent immersion in 43°C water considerably, and they discussed the implications to heat quarantine treatment research. They note that these types of heat pre-treatments are commonly and inadvertently applied to fruits by the sun and warm packing sheds, for example. Modification of insect mortality levels by pre-treatment temperature changes may also threaten cold quarantine treatments (see Chapter 3). Because these effects seem to apply across all taxonomic groups and may result from relatively mild changes in temperature, and given that pre-treatment temperature changes on a commercial scale are often unavoidable, we wonder why this has not lead to notable failures in temperature quarantine treatments. Part of the reason may be that quarantine security is generally very conservative (Vail et al. 1993). This could become a problem if quarantine security is liberalized (Liquido et al. 1997). But also, it may be that when live larvae are found in a temperature-treated fruit shipment other reasons are blamed.

Because of the significant potential for failure of a quarantine treatment caused by pre-treatment temperature variation, it is recommended that researchers incorporate reasonable pre-treatment temperature modification into research designed to develop temperature-based quarantine treatments. This has not been done intentionally for any confirmatory research on temperature quarantine treatments; on the contrary, researchers usually prefer to maintain pre-treatment temperatures at constant, mild levels.

Most Tolerant Stage

Different insect stages of the same species often vary in susceptibility to heat and cold. A successful quarantine treatment must achieve the required level of mortality against the most tolerant stage that may be present within the commodity. With some insects, such as most weevils, any stage from egg through the adult may be present. With tephritid fruit flies, eggs and larvae may be present. However, cold treatments are usually applied to the commodity after it is packed, and it may be a couple of days before the entire load cools down to temperatures which will halt development. In some circumstances or with some commodities it is entirely possible that they may sit at ambient temperature for a while before being put in cold storage, however unfavorable that may be for commodity quality. Large shipments of citrus may take several days to cool after they are loaded. Therefore, it is possible that larvae could emerge and be in the postfeeding, prepupal, or pupal stages in the cartons before the treatment begins. Hallman (1998) observed that Mexican fruit flies increased greatly in tolerance to 1.1°C on the third day after pupariation, when the phanerocephalic pupal stage formed. Stages before that period were equal to or less tolerant of cold than feeding third instars.

When searching for the stage most tolerance to a treatment, the location of the different stages and the slope of the dose-mortality relationship are important concerns. For example, latter instars will normally be deeper inside a fruit and more protected from rapid changes in temperature which may allow for some conditioning to occur. Insects near the surface of the fruit are subject to more total lethal heating than those near the center, as heating curves of grapefruits demonstrate (Fig. 8.2). Although one stage may be more tolerant than another at median ranges of mortality, at high levels of mortality, which are most important in quarantine treatments, the two stages may reverse relative susceptibility. For example, Heather et al. (1997) found that Queensland and Mediterranean fruit fly eggs were more tolerant of hot air than last instars as the temperature rose to 43-44°C, but at 45°C (>98% mortality), last instars were more tolerant.

Variability of Commodity

A two-stage hot water immersion quarantine treatment of Hawaiian papayas was developed but failed after a few years. Papayas not more than one-quarter ripe were immersed in 42°C water for 30 min followed by 49°C water for 20 min and finally hydrocooled with ambient water spray (Couey & Hayes 1986). This treatment was designed to kill only eggs and early instar larvae near the fruit surface. Papayas more than one-quarter ripe might have larvae in the center, which might not be killed by the treatment. However, in March of 1987 nine of 16,000 papayas which were subjected to the two-stage immersion treatment were found to be infested with oriental fruit fly, *Bactrocera dorsalis*, larvae (Zee et al. 1989). It was discovered that some papayas had an opening at the blossom end which allowed for fruit flies to oviposit in the papaya at an earlier stage and for larvae to more quickly enter the center of the fruit where the lowered treatment severity could allow for survivors. A survey revealed that papayas from 5-31% of papaya trees from commercial orchards in the Puna district of Hawaii in 1987 had that opening to varying degrees. Zee et al. (1989) cited P. J. Ito and H. Y. Nakasone who said that the defect was not present in the original commercial papaya cultivars but must represent inadvertent cross pollination of these highly self-pollinated hermaphrodite inbreds. In any case, the defect must have been present to some degree when the research on the two-stage papaya treatment was done just a few years before the defect was first encountered. Couey & Hayes (1986) did note six survivors of the treatment during the research phase, but dismissed them as "accidental reinfestation". Natural infestation studies were done with fruits taken from only two orchards. The large scale confirmatory tests (80,000 fruits) were conducted entirely with papayas from one. That grower may have simply had a low incidence of the blossom end defect and/or a low incidence of fruit fly infested papayas.

The lesson to be learned from this incident is that fruits used in quarantine research should be sampled from a wide range of the potential growing area to increase chances that fruits which may yield drastically different responses would be represented. Also, researchers should not lightly dismiss unexpected anomalies in research results.

Commercial-Scale Engineering Considerations

Although a temperature treatment can be devised and function efficiently in a small scale research facility (Gaffney 1990, Gaffney & Armstrong 1990, Gaffney et al. 1990), it may not operate as expected commercially because of problems related to scale (Williamson & Winkleman 1994). These problems are more acute for heat treatments than cold treatments because heat treatments are short. Also, they are more critical for heated air treatments than for heated water treatments because of several properties of air: (1) Heated air takes the path of least resistance in a fruit load and may leave areas with lower temperatures (Fig. 8.3). (2) As air moves through a fruit load the temperature is reduced and the humidity either increases or decreases. (3) Air speed is directly proportional to fruit heating rate and inversely related to density of fruit load. Williamson & Winkleman (1989) discuss challenges in developing a commercial-scale heated air treatment facility. The basic challenge is coordinating economic,

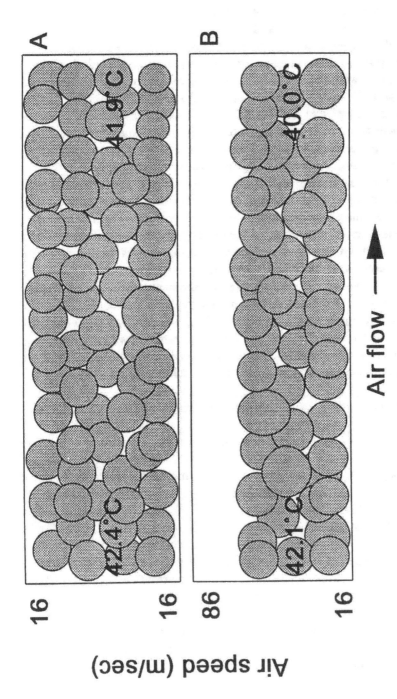

FIGURE 8.3 Effect of 7.5 cm air space (bottom diagram) on air speed and center temperatures of grapefruits after 160 min at 43.3°C, 100% relative humidity; container is 2 m long (Hallman, unpublished data).

engineering, and biological factors in an environment that is extremely restrictive in the tolerance of treatment variation. An additional difficulty concerns the insufficient biological knowledge of tolerances of the commodity to the treatment, especially as it relates to handling of the commodity before and after treatment. In this regard, it would be very beneficial if an engineer experienced in the heating and cooling of fresh commodities were involved in quarantine treatment research once it approaches the commercial phase.

Recent Advances in Temperature Treatments

Considerable research on temperature quarantine treatments, especially heat, has been conducted as alternatives to methyl bromide and as improvements to market quality. Recent developments that may have not been covered sufficiently in other reviews are discussed in the following sections.

Preconditioning to Avoid Chilling Injury

Most of the current developments in quarantine treatments based on temperature involve heat. Nonetheless, there are novel developments in cold treatments. Chilling injury to fruits has limited the use of cold as a quarantine treatment for many commodities. Paull & McDonald (1994) review techniques that have been used to reduce chilling injury, such as holding fruits at temperatures several degrees above the quarantine treatment temperature for several days prior to cold treatment or heating fruit to above ambient temperatures for shorter periods of time before treatment. For example, Jessup (1991, 1994) showed that Hass avocados could withstand insecticidal cold treatments if preconditioned with a short hot water dip similar to that used for fungicidal treatment. However, any pre-treatment which favors tolerance of the fruit to a cold treatment can also be suspected of favoring survival of the pest inside of the fruit (see Chapter 3 for examples of insect cold hardening). Before pre-conditioning procedures touted to better maintain commodity quality after a cold quarantine treatment are offered to processors they should be checked to determine if they likewise improve pest survival.

Rapid Cooling

As with many treatments, efficacy of cold treatment may be positively related to treatment rate. Gould & Hennessey (1997) found a faster rate of mortality of Caribbean fruit fly, *Anastrepha suspensa*, large third instars in carambolas cooled in 40-45 min to a holding temperature of 1.1°C versus those cooled to 1.1°C in over 24 h. Although rapid cooling may not be commercialy feasible, it may indicate a problem when cold treatments are applied on a commercial scale. The faster cooling rate may have prevented cold hardening by the insect. Because the cooling rate of fruit in large, commercial coolers is slower than that usually achieved in small research coolers, it is possible that a cold treatment applied on a commercial scale may have lower mortality than that same treatment when performed on a research scale. If that proves to be a concern, then the cooling rate during research should not be faster than that possible commercially.

Preconditioning to Avoid Heat Injury

As with cold, damage to treated produce has been a concern with heat treatments. Pre-treatment heating at a lower temperature than that used to kill insects has been used to ameliorate damage caused by the subsequent heat treatment (Paull & McDonald 1994). Almost 60 years ago Jones (1940) found that certain fruits such as papaya and eggplant tolerated insecticidal vapor heat treatments better if held between 38 and 43°C for eight hours prior to treatment. This increased tolerance to temperature stress is likely due to the production of stress related proteins and will probably occur in any insects in the commodities, also. Therefore, as with rapid cooling, before any treatment using a pre-treatment designed to increase commodity tolerance to temperature extremes is offered for commercial use, it must be tested for its effect on insect survival.

Post-Heat Treatment Cooling to Lessen Injury

Likewise, rapid post-heating cooling has been postulated as a means of reducing heat damage to produce, although we could not find much actual data on the subject. Shellie (1994) found that rapid post-heating cooling did not preserve quality of 'Manila' mangoes.

Rapid cooling of fruit after a heat treatment below temperatures lethal to insects might be expected to increase the risk of insects surviving.

However, results have been inconsistent. Hallman & Sharp (1990a) found higher mean estimates of time needed to obtain 99.9968% mortality of Caribbean fruit fly third instars in mangoes immersed in 46.1-46.7°C water when the mangoes were cooled to near 40°C in 10 min versus about 35 min, althought the 95% fiducial limits overlapped. In a larger scale follow-up test one larva from an estimated 17,589 larvae survived the rapid cooling procedure while none survived the slow cooling approach. In a test of Caribbean fruit fly-infested carambolas dipped in 46-46.4°C water for 45 min, immersion in 13°C water immediately after heating did not result in any survivors of an estimated 30,126 larvae (Hallman & Sharp 1990b). However, the heat treatment was probably excessive, as there were no survivors of an estimated 190,687 larvae treated and air cooled, which is double the number needed to achieve the commonly accepted level of quarantine security. Although probit-analysis estimates of $LT_{99.9968}$ needed for 46-46.3°C vapor heat treatment were very similar regardless of whether Caribbean fruit fly infested carambolas were cooled in 13°C water or left in ambient air after heating, no larvae of an estimated 95,327 survived the air-cooled carambolas while three of an estimated 45,856 survived in the water-cooled fruits after treatment with vapor heat for 90 min (Hallman 1990b). In a fourth study, there was no significant difference in preliminary tests of Caribbean fruit fly larval mortality among rapid and slow cooling of heat-treated guavas (Gould & Sharp 1992). In a fifth study, Sharp & Gould (1994) found that estimates of time required to produce 99.9968% mortality to heat-treated Caribbean fruit fly immatures in grapefruits were significantly longer (based on non-overlapping 95% fiducial limits) for grapefruits cooled slowly in ambient air than those cooled relatively faster in 10°C water. However, no large-scale test confirming this observation was performed, and the data did not fit the analysis (probit) used. Sharp & Gould (1994) concluded that rapid water cooling of heated grapefruits would not jeopardize quarantine security. In their discussion to generalize this conclusion, they erroneously cited Hallman & Sharp (1990a) by stating that "lethal temperatures [were] not reached" because mango interior temperatures were <42°C when cooling was initiated. In fact, temperatures were about 43.3°C when cooling was initiated, and no larvae of an estimated 17,589 survived the slow-cooled treatment, which was certainly lethal. Small temperature differences in the range of 43-46° may cause significant differences in fruit fly mortality (Jang 1991). Furthermore, interior temperatures in slow-cooled mangoes continued to rise for over five minutes after removal from the heat source, whereas interior temperatures in faster-cooled mangoes began to drop immediately after removal from the heat source (Hallman & Sharp 1990a).

It is our conclusion after reviewing the five studies on the subject that the possibility that rapid cooling of heated produce may lower insect mortality is significant enough to warrant its testing with large numbers of insects, not simply mortality estimates, before allowing any accelerated cooling of heated produce to be performed commercially. As an example, this approach was followed by Armstrong et al. (1989, 1995a).

Radio Frequency Heating

Commercial heat quarantine treatments warm the surface of the commodity and then rely on conduction to heat the interior. Some research has been carried out on radio frequency (including microwave) heating of fresh commodties without much success (Hallman & Sharp 1994). Nonthermal detrimental effects of radio waves on organisms have been repeatedly postulated but have not been convincingly demonstrated (Nelson 1995). The hypothetical advantages of radio frequency heating over conductive heating (uniform heating of whole fruit and selective heating of insects inside of fruits) have not materialized. No new developments in radio frequency treatment offer promise for its use as a quarantine treatment against pests of fresh produce. Nevertheless, any promising research leads in this area should be explored.

Ohmic Heating

Ohmic heating is a non-conductive thermal technique which may show promise as a quarantine treatment. Considerable research and substantial commercial application have been done with ohmic heating as a means of sterilzing foodstuffs that is less damaging to the nutritive quality of the food than alternative heating methods (Zoltai & Swearingen 1996, Giese 1996). In ohmic heating, an object heats when an electrical current is passed through it due to the object's electrical resistance in the same fashion that an electric stove element heats. We have done preliminary work with ohmic heating as a quarantine treatment and have found that it is possible to heat a small batch of whole fruits rapidly and kill fruit fly larvae inside of fruits when the fruits are immersed in a dilute salt solution (Hallman & S. Sastry, unpublished data). Ohmic heating may heat fruits more uniformily than radio frequency heating.

Heated Modified Atmospheres

Modified atmospheres, generally low O_2 and/or high CO_2, can be used to disinfest fresh horticultural commodities of insect pests (Carpenter & Potter 1994, Hallman 1994a). Much of the earlier work on modified atmosphere quarantine treatments was done at low temperatures. Although low temperature-modified atmosphere quarantine treatments are possible, no synergistic or additive effect seems to occur. Often, it was not determined if the modified atmosphere exerted any effect beyond that caused simply by the cold temperature. An exception is Shelton et al. (1996) who found significantly greater mortality of western flower thrips, *Frankliniella occidentalis*, and green peach aphid, *Myzus persicae*, when stored at 2°C for 4 days in an atmosphere containing 0.1% O_2 versus air.

The successful combination of heat with modified atmospheres is a more recent development. Time to 100% mortality of several insects affecting raisins was reduced by half when the temperature in controlled atmosphere storage was raised from 16 to 27°C (Soderstrom & Brandl 1984). Likewise, as temperature increased between 15.6 and 26.7°C, the LT_{95} for two stored-product moths in the presence of 0.5-5% O_2 and 10% CO_2 was reduced by about one-half to two-thirds (Soderstrom et al. 1986). Whiting et al. (1991, 1992, 1995, 1996) studied the mortality responses of eggs and larvae of six tortricid species to modified atmospheres between 12.5-40°C. Estimated LT_{99}s were reduced from 100-250 h at the low temperatures to just a few hours at 40°C. Exposure of red flour beetle, *Tribolium castaneum*, to 38°C for 24-48 h before a combined 38°C-modified atmosphere treatment reduced the effectiveness of the combined treatment (Soderstrom et al. 1992). However, the opposite effect was observed with diapausing twospotted spider mite, *Tetranychus urticae* (Whiting & van den Heuvel 1995). Pretreatment at 40°C for 8-24 h decreased tolerance of the mites to a subsequent combined 40°C-modified atmosphere treatment. LT_{95}s for diapausing codling moth larvae in walnuts heated at 39-45°C were much shorter in atmospheres containing 0.5% O_2 and especially 98% CO_2 compared with air (Soderstrom et al. 1996a, 1996b). Similarly, heating codling moth fifth instars in cherries under an atmosphere comprising 1% O_2 and 15% CO_2 resulted in greatly reduced LT_{99}s compared with heating the insects in cherries in air (Neven & Mitcham 1996). Shellie et al. (1997) achieved 100% Mexican fruit fly larval mortality in grapefruits in an atmosphere containing 1% O_2 at 46°C in 3.5 h versus 5 h for heating in air.

Another novel approach combined cold as a second treatment following heat and modified atmospheres. Chervin et al. (1997) found that three lepidopterans infesting pears in Australia were killed by 30 h at 30°C in an atmosphere consisting of 0.3 ± 0.2 kPa O_2 plus 4.6 ± 0.9 kPa CO_2 followed by one month at 0°C; pears tolerated the treatment. However, one is left wondering what effect the 30°C-modified atmosphere by itself would have had on insect mortality; perhaps it would suffice without the one month cold storage.

Fruit coatings modify internal fruit atmospheres, and the combination of coating with heat increased fruit fly mortality over heat alone (Fig. 8.4).

In summary, the combination of heat with controlled atmospheres is very favorable to killing insects. However, fruits have often suffered from the treatment (Shellie et al. 1997, Neven & Mitcham 1996). The challenge is to devise treatments which will kill quarantined pests without harming fruit quality.

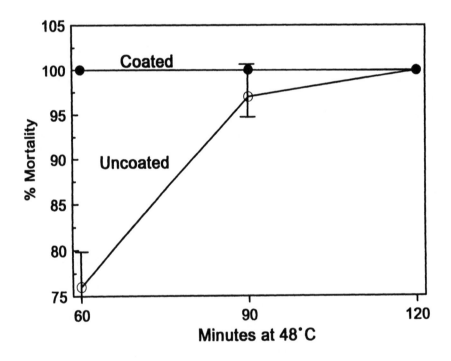

FIGURE 8.4 Mortality of Caribbean fruit fly last instars in coated (Nature Seal containing 2% methyl cellulose and 10% food-grade shellac) and uncoated grapefruits subjected to 48°C air (Hallman et al. 1994).

Generic Heat Treatment

It has been proposed that fruit flies can be killed when a commodity is heated at a high enough temperature and for sufficient time so that the center reaches a minimum of 47°C (McGuire & Sharp 1997). We do not see the wisdom to this approach. Many fruits cannot tolerate that degree of heat, and many of those that can will still suffer loss in quality after such an exposure (McDonald & Miller 1994, Paull & McDonald 1994). Most fruit flies can be killed at lower temperatures (Sharp 1994, Hallman & Armstrong 1994). Although, in some cases, this target temperature might not suffice to kill tephritid immatures (Table 8.1). Also, insect mortality is related to heating rate, not simply target temperature (Neven 1998).

Killing insects is only one part of a quarantine treatment. Conserving fruit quality is just as important for the development of a viable treatment. The simplicity offered by generic heat treatments is more than offset by the additional cost of overtreatment and possible loss in commodity quality in those cases where a milder treatment would suffice. We believe that researchers should always attempt to devise a treatment which, while assuring quarantine security, causes the least damage to the commodity, ease and cost of application being other important concerns.

Sequential Heat and Cold

Heat may increase the tolerance of some fruits to cold storage (Jessup 1991, McDonald & Miller 1994). Three studies examined the effect of a heat treatment followed by cold storage on fruit infesting insects. Gould (1988) estimated that immersion of grapefruits in 43.3°C water for 100 min followed by 7 days at 1.1°C would provide 99.9968% mortality of Caribbean fruit fly last instars. Heat alone would have required an estimated 175 min, and to kill Caribbean fruit flies with 1.1°C alone requires at least 12 days. Neven (1994) likewise treated codling moth last instars with 43-45°C heat before subjecting them to 0 or 5°C storage. Cold storage by itself up to 28 days did not affect mortality, but when it followed 30-min heat treatments at 44 or 45°C, mortality increased to a plateau level under 100% after 7 days of cold storage. However, heat may also increase tolerance of insects to subsequent cold (see Chapters 2 and 3). Dentener et al. (1997) studied a sequential heat-cold treatment to disinfest persimmons of two insects, one of which was more susceptible to cold while the other was more susceptible to heat.

TABLE 8.1. Minimum interior fruit temperatures to achieve quarantine security (usually 99.9968%mortality) for various heat treatments

Fruit	Insect	Treatment	Min. interior temp. (°C)	Reference
Carambola	Caribbean fruit fly	Air at 47°C	45.5	Sharp & Hallman (1992)
Grapefruit	Caribbean fruit fly	Air at 48°C	44	Sharp (1993)
Grapefruit	Mexican fruit fly	Air at 50°C	48	Mangan & Ingle (1994)
Mango	Caribbean fruit fly	Air at 48°C	46.1	Sharp (1992)
Mango	Mexican fruit fly	Air at 50°C	48	Mangan & Ingle (1992)
Orange	Caribbean fruit fly	Air at 48°C	44	Sharp & McGuire (1996)
Papaya	Three fruit flies	Air at 48.5°C	47.2	Armstrong et al. (1989)

If heat followed by cold (or vice versa) is to be considered as a quarantine treatment, then the optimum combination of temperatures and times that favor the commodity while harming the pest must be sought. Seemingly minor variations of the treatment could have very different outcomes. For example, while a positive effect on codling moth mortality was observed after 30 min at 44°C followed by cold storage, no effect was observed when the heating was done at one degree lower (Neven 1994).

Temperature Treatment of Cut Flowers and Foliage

Cold and heat might become effective substitutes for methyl bromide fumigation of floral products in some cases. The recommended storage temperatures of many common cut floral products, such as carnations, chrysanthemums, roses, and ferns (Hardenburg et al. 1986), are within the range needed for control of many insects. Hansen & Hara (1994) reviewed research performed on temperature disinfestation treatments of tropical cut flowers and noted some that might tolerate an insecticidal cold treatment. Unfortunately, the cold temperature-mortality relationship for quarantined pests on cut floral products has hardly been investigated. An inexhaustive review on our part resulted in only two papers (Seaton & Joyce 1989, 1993), and surrogate insects, not actual flower pests, were used.

Heated water and air treatments have been investigated to disinfest cut flowers of pests (Hansen & Hara 1994). Heat seems to work well for some tropical flowers, such as heliconia and bird of paradise. Hara et al. (1994) achieved ≥99.7% mortality of green scale, *Coccus viridis*, on cape jasmine propagative cuttings with a 10 min dip in 49°C water without damaging the ability of the cuttings to grow. They felt that this level of mortality coupled with management practices against the pest in the field would be sufficient to ensure quarantine security in exported cape jasmine cuttings. A hot water treatment (12-15 min at 49°C) plus field management of several pests of red ginger flowers was also believed to provide export-level quarantine security (Hara et al. 1996).

Future Directions

Temperature treatments comprise the largest area of research into quarantine treatments. Nevertheless, there are many avenues which merit further exploration. Hot water immersion of mangoes has been one of the greatest successes of a temperature quarantine treatment (Sharp 1994), and the technique is simple, economical, secure from rapid temperature changes, and applicable to many potential problems that may affect other temperature treatments (Hallman & Quinlan 1996). However, research on hot water immersion seems to have taken a back seat to heated air due, in some part, to problems with commodity quality. However, McGuire (1991) found that damage to mangoes could be reduced greatly if the fruits were immersed in water that was gradually heated from 22 to 48°C until the mango centers reached 44°C rather than being immediately immersed in 46°C water. Other variations in the standard hot-water immersion technique should be tried, such as bubbling air through the water to possibly avoid damage thought to be caused by anoxia.

For the most part there has not been a clear benefit to insect mortality from combining cold with anoxia in a quarantine treatment. However, Yocum & Denlinger (1994) found that anoxia blocked rapid cold hardening in pharate adult *Sarcophaga crassipalpis* exposed to 0°C for 2 h followed by -10°C for 2 h. This finding may hint at heretofore unknown opportunities for combining anoxia and low temperatures as quarantine treatments.

Heat has been shown to combine well with modified atmospheres, cold is combined with methyl bromide fumigation on a commercial scale, and combinations of hot water immersion and ethylene dibromide fumigation have been studied (Mangan & Sharp 1994). Other combinations involving temperatures should be investigated. Combination treatments may allow fruits which cannot tolerate higher doses of either treatment to be treated efficaciously.

The issues of conditioning of insects to temperature extremes and stress proteins present serious theoretical consequences for temperature quarantine treatments which have not been confronted at the regulatory or commercial levels. Live insects have been found after approved temperature quarantine treatments that have supposedly been applied correctly. Sometimes the reason is clearly determined, as with the papaya blossom end defect (Zee et al. 1989). Other times it has simply been concluded that the treatment was not applied correctly, although proof was not certain. Efforts to reduce the severity of quarantine treatments may result in this problem becoming clearly manifest, or maybe the problem is not serious in a practical sense. In any case, these issues need to be examined in the commercial setting, because they have obviously been shown to exist in the laboratory.

References

Anonymous. 1907. Cool storage and fruit fly. J. Dept. Agr., Western Australia 15: 252-253.

Anonymous. 1993. Protection of stratospheric ozone. Federal Register 58: 65018-65082.

Armstrong, J. W., J. D. Hansen, B. K. S. Hu & S. A. Brown. 1989. High-temperature, forced-air quarantine treatment for papayas infested with tephritid fruit flies (Diptera: Tephritidae). J. Econ. Entomol. 82: 1667-1674.

Armstrong, J. W., B. K. S. Hu & S. A. Brown. 1995a. Single-temperature forced hot-air quarantine treatment to control fruit flies (Diptera: Tephritidae) in papaya. J. Econ. Entomol. 88: 678-682.

Armstrong, J. W., S. T. Silva & V. M. Shishido. 1995b. Quarantine cold treatment for Hawaiian carambola fruit infested with Mediterranean fruit fly, melon fly, or oriental fruit fly (Diptera: Tephritidae) eggs and larvae. J. Econ. Entomol. 88: 683-687.

Baker, A. C. 1952. The vapor-heat process, *Insects: The Yearbook of Agriculture*, pp 401-404. Dept. of Agr. U.S. Govt. Printing Office. Washington, D. C.

Beckett, S. J. & D. E. Evans. 1997. The effects of thermal acclimation on immature mortality in the Queensland fruit fly *Bactrocera tryoni* and the light brown apple moth *Epiphyas postvittana* at a lethal temperature. Entomologia Experimentalis et Applicata 82: 45-51.

Buchanan, A. 1994. Tropical fruits: the social, political and economic issues. *In:* B. R. Champ, E. Highley & G. I. Johnson, eds. *Post Harvest Handling of Tropical Fruits.* pp 18-26. Australian Centre for Internat. Agric. Res Proc. No. 50. Conference held at Chiang Mai, Thailand, 19-23 July 1993.

Carpenter, A. & M. Potter. 1994. Controlled atmospheres, *In:* J. L. Sharp & G. J. Hallman, eds. *Quarantine Treatments for Pests of Food Plants.* pp. 171-198. Westview Press, Boulder, Colorado.

Chambers, D. L. 1977. Quality control in mass rearing. Annual Rev. Entomol. 22: 289-308.

Chervin, C., S. Kulkarni, S. Kreidl, F. Birrell & D. Glenn. 1997. A high temperature/low oxygen pulse improves cold storage disinfestation. Postharvest Biology and Technol. 10: 239-245.

Couey, H. M. & C. F. Hayes. 1986. A quarantine system for papayas using fruit selection and a two-stage hot-water treatment. J. Econ. Entomol. 79: 1307-1314.

Dentener, P. R., K. V. Bennett, L. E. Hoy, S. E. Lewthwaite, P. J. Lester, J. H. Maindonald & P. G. Connolly. 1997. Postharvest disnifestation of lightbrown apple moth and longtailed mealybug on persimmons using heat and cold. Postharvest Biol. and Technol. 12: 255-264.

Fuller, C. 1906. Cold storage as a factor in the spread of insect pests. Natal Agricultural J. 9: 656.

Gaffney, J. J. 1990. Warm air/vapor heat research facility for heating fruits for insect quarantine treatments. Paper No. 906615, American Soc. of Agric. Engineers International Winter Meeting, Chicago, Illinois.

Gaffney, J. J., & J. W. Armstrong. 1990. High-temperature forced-air research facility for heating fruits for insect quarantine treatments. J. Econ. Entomol. 83: 1959-1964.

Gaffney, J. J., G. J. Hallman & J. L. Sharp. 1990. Vapor heat research unit for insect quarantine treatments. J. Econ. Entomol. 83: 1965-1971.

Giese, J. 1996. Commercial development of ohmic heating garners 1996 Industrial Achievement Award. Food Technol. 50: 114-115.

Gould, W. P. 1988. A hot water/cold storage quarantine treatment for grapefruit infested with the Caribbean fruit fly. Proc. Fla. State Hort. Soc. 101: 190-192.

Gould, W. P. 1994. Cold Storage, *In*: J. L. Sharp & G. J. Hallman, eds. *Quarantine Treatments for Pests of Food Plants*. pp 119-132. Westview Press, Boulder, Colorado.

Gould, W. P. & M. K. Hennessey. 1997. Mortality of *Anastrepha suspensa* (Diptera: Tephritidae) in carambola treated with cold water precooling and cold storage. Florida Entomol. 80: 79-84.

Gould, W. P. & J. L. Sharp. 1990. Cold-storage quarantine treatment for carambolas infested with the Caribbean fruit fly (Diptera: Tephritidae). J. Econ. Entomol. 83: 458-460.

Gould, W. P. & J. L. Sharp. 1992. Hot-water immersion quarantine treatment for guavas infested with Caribbean fruit fly (Diptera: Tephritidae). J. Econ. Entomol. 85: 1235-1239.

Hallman, G. J. 1990a. Survival and reproduction of Caribbean fruit fly (Diptera: Tephritidae) adults immersed in hot water as third instars. J. Econ. Entomol. 83: 2331-2334.

Hallman, G. J. 1990b. Vapor-heat treatment of carambolas infested with Caribbean fruit fly (Diptera: Tephritidae). J. Econ. Entomol. 83: 2340-2342.

Hallman, G. J. 1994a. Controlled atmospheres, *In* R. E. Paull & J. W. Armstrong, eds. *Insect Pests and Fresh Horticultural Products: Treatments and Responses*. pp 121-136. CAB International, Wallingford, UK.

Hallman, G. J. 1994b. Mortality of third-instar Caribbean fruit fly (Diptera: Tephritidae) reared at three temperatures and exposed to hot water immersion or cold storage. J. Econ. Entomol. 87: 405-408.

Hallman, G. J. 1996. Mortality of third instar Caribbean fruit fly (Diptera: Tephritidae) reared in diet or grapefruits and immersed in heated water or grapefruit juice. Florida Entomol. 79: 168-172.

Hallman, G. J. 1998. Lethality of cold to Mexican fruit fly (Diptera: Tephritidae) third instars and pupae. J. Econ. Entomol. 91: (in press)

Hallman, G. J. & J. W. Armstrong. 1994. Heated air treatments, *In* J. L. Sharp & G. J. Hallman, eds. *Quarantine Treatments for Pests of Food Plants*, pp 149-163. Westview Press, Boulder, Colorado.

Hallman, G. J., J. J. Gaffney & J. L. Sharp. 1990. Vapor heat treatment for grapefruit infested with Caribbean fruit fly (Diptera: Tephritidae). J. Econ. Entomol. 83: 1475-1478.

Hallman, G. J., M. O. Nisperos-Carriedo, E. A. Baldwin & C. A. Campbell. 1994. Mortality of Caribbean fruit fly (Diptera: Tephritidae) immatures in coated fruits. J. Econ. Entomol. 87: 752-757.

Hallman, G. J. & M. Quinlan. 1996. Synopsis of postharvest quarantine research, *In* B. A. McPheron & G. J. Steck, eds. *Fruit Fly Pests: A World Assessment of Their Biology and Management.* pp 473-477. St. Lucie Press, Delray Beach, Florida.

Hallman, G. J. & J. L. Sharp. 1990a. Mortality of Caribbean fruit fly (Diptera: Tephritidae) larvae infesting mangoes subjected to hot-water treatment, then immersion cooling. J. Econ. Entomol. 83: 2320-2323.

Hallman, G. J. & J. L. Sharp. 1990b. Hot-water immersion quarantine treatment for carambolas infested with Caribbean fruit fly (Diptera: Tephritidae). J. Econ. Entomol. 83: 1471-1474.

Hallman, G. J. & J. L. Sharp. 1994. Radio frequency heat treatments, *In* J. L. Sharp & G. J. Hallman, eds. *Quarantine Treatments for Pests of Food Plants.* pp 165-170. Westview Press, Boulder, Colorado.

Hansen, J. D., J. W. Armstrong, B. K. S. Hu & S. A. Brown. 1990. Thermal death of oriental fruit fly (Diptera: Tephritidae) third instars in developing quarantine treatments for papayas. J. Econ. Entomol. 83: 160-167.

Hansen, J. D. & A. H. Hara. 1994. A review of postharvest disinfestation of cut flowers and foliage with special reference to tropicals. Postharvest Biol. and Technol. 4: 193-212.

Hansen, J. D. & J. L. Sharp. 1997. Thermal death in third instars of the Caribbean fruit fly (Diptera: Tephritidae): density relationships. J. Econ. Entomol. 90: 540-545.

Hara, A. H., T. Y. Hata, B. K. S. Hu, R. T. Kaneko & V. L. Tenbrink. 1994. Hot-water immersion of cape jasmine cuttings for disinfestation of green scale (Homoptera: Coccidae). J. Econ. Entomol. 87: 1569-1573.

Hara, A. H., T. Y. Hata, V. L. Tenbrink, B. K.-S. Hu & R. T. Kaneko. 1996. Postharvest heat treatment of red ginger flowers as a possible alternative to chemical insecticidal dip. Postharvest Biol. and Technol. 7: 137-144.

Hardenburg, R. E., A. E. Watada & C. Y. Wang. 1986. *The Commercial Storage of Fruits, Vegetables, and Florist and Nursery Stocks.* U.S. Dept. Agr., Agric. Res. Serv. Handbook No. 66.

Heather, N. W., R. J. Corcoran & R. A. Kopittke. 1997. Hot air disinfestation of Australian 'Kensington' mangoes against two fruit flies (Diptera: Tephritidae). Postharvest Biol. and Technol. 10: 99-105.

Heather, N. W., L. Whitfort, R. L. McLaughlan & R. Kopittke. 1996. Cold disinfestation of Australian mandarins against Queensland fruit fly (Diptera: Tephritidae). Postharvest Biol. and Technol. 8: 307-315.

Jang, E. B. 1991. Thermal death kinetics and heat tolerance in early and late third instars of the oriental fruit fly (Diptera: Tephritidae). J. Econ. Entomol. 84: 1298-1303.

Jessup, A. J. 1991. High temperature dip and low temperatures for storage and disinfestation of avocados. HortSci. 26:1420.

Jessup, A. J. 1994. Quarantine disinfestation of 'Hass' avocados against *Bactrocera tryoni* (Diptera: Tephritidae) with a hot fungicide dip followed by cold storage. J. Econ. Entomol. 87: 127-130.

Jones, W. W. 1939. The influence of relative humidity on the respiration of papaya at high temperatures. Proc. American Soc. Hort. Sci. 37: 119-124.

Jones, W. W. 1940. Vapor-treatment for fruits and vegetables grown in Hawaii. Circular No. 16, Hawaii Agr. Expt. Sta. Honolulu. 8 pp.

Jones, W. W., J. J. Holzman & A. G. Galloway. 1939. The effect of high-temperature sterilization on the solo papaya. Hawaii Agric. Exper. Sta. Circular No. 14, Honolulu. 8 pp.

Kelly, M. 1996. Methyl bromide--fixer, ogre or an opportunity? International Food Hygiene 7: 23, 25, 27.

Lester, P. J. & D. R. Greenwood. 1997. Pretreatment induced thermotolerance in lightbrown apple moth (Lepidoptera: Tortricidae) and associated induction of heat shock protein synthesis. J. Econ. Entomol. 90: 199-204.

Liquido, N. J., R. L. Griffin & K. W. Vick, eds. 1997. *Quarantine Security for Commodities: Current Approaches and Potential Strategies.* U. S. Dept. Agr., Agric. Res. Service.

Mangan, R. L. & S. J. Ingle. 1992. Forced hot-air quarantine treatment for mangoes infested with West Indian fruit fly (Diptera: Tephritidae). J. Econ. Entomol. 85: 1859-1864.

Mangan, R. L. & S. J. Ingle. 1994. Forced hot-air quarantine treatment for grapefruit infested with Mexican fruit fly (Diptera: Tephritidae). J. Econ. Entomol. 87: 1574-1579.

Mangan, R. L. & J. L. Sharp. 1994. Combination and multiple treatments, *In* J. L. Sharp & G. J. Hallman, eds. *Quarantine Treatments for Pests of Food Plants.* pp 239-247. Westview Press, Boulder, Colorado.

McDonald, R. E. & W. R. Miller. 1994. Quality and Condition Maintenance, *In* J. L. Sharp & G. J. Hallman, eds. *Quarantine Treatments for Pests of Food Plants.* pp 249-277. Westview Press, Boulder, Colorado.

McGuire, R. G. 1991. Concomitant decay reductions when mangoes are treated with heat to control infestations of Caribbean fruit flies. Plant Disease 75: 946-949.

McGuire, R. G. & J. L. Sharp. 1997. Quality of colossal Puerto Rican 'Keitt' mangoes after quarantine treatment in water at 48°C. Tropical Science 37: 154-159.

Minnis, D. C. 1994. Prospects for marketing tropical fruits in Asia. *In* B. R. Champ, E. Highley & G. I. Johnson, eds. *Post Harvest Handling of Tropical*

Fruits. pp 56-64. ACIAR Proc. No. 50. Conf. held at Chiang Mai, Thailand, 19-23 July 1993.

Nelson, S. O. 1995. Assessment of RF and microwave electric energy for stored-grain insect control. Amer. Soc. Agric. Engin. Paper No. 956527 presented at the June 18-23, 1995 Amer. Soc. Agric. Engin. Ann. Internat. Meeting in Chicago, Illinois.

Neven, L. G. 1994. Combined heat treatments and cold storage effects on mortality of fifth-instar codling moth (Lepidoptera: Tortricidae). J. Econ. Entomol. 87: 1262-1265.

Neven, L. G. 1998. Effects of heating rate on the mortality of fifth-instar codling moth (Lepidoptera: Tortricidae). J. Econ. Entomol. 91: 297-301.

Neven, L. G. & E. J. Mitcham. 1996. CATTS (controlled atmosphere/temperature treatment system): a novel tool for the development of quarantine treatments. American Entomol. 42: 56-59.

Paull, R. E. & R. E. McDonald. 1994. Heat and cold treatments, *In* R. E. Paull & J. W. Armstrong, eds. *Insect Pests and Fresh Horticultural Products: Treatments and Responses*. pp 191-222. CAB International, Wallingford, United Kingdom.

Proctor, F. J. & J. P. Cropley. 1994. Trends and changes in the European market for tropical fruits and their impact on technology requirements. *In* B. R. Champ, E. Highley and G. I. Johnson, eds. *Post Harvest Handling of Tropical Fruits*. pp 65-72. ACIAR Proceedings No. 50. Conference held at Chiang Mai, Thailand, 19-23 July 1993.

Richardson, H. H. 1952. Cold treatment of fruits. *Insects: The Yearbook of Agriculture*. pp 404-406. Dept. of Agr. U. S. Govt. Printing Office. Washington, D. C.

Ruckelshaus, W. D. 1984. Ethylene dibromide; amendment of notice of intent to cancel registrations of pesticide products containing ethylene dibromide. Federal Register 49:14182-14185.

Seaton, K. A. & D. C. Joyce. 1989. Postharvest disinfestation of cut flowers for export. *Horticultural Research and Extension Update-1989*, Western Australia Dept. Agr., Hort. Div. Workshop, June 19-20, Perth.

Seaton, K. A. & D. C. Joyce. 1993. Effects of low temperature and elevated CO_2 treatments and of heat treatments for insect disinfestation on some native Australian cut flowers. Scientia Horticulturae 56: 119-133.

Shannon, M. J. 1994. APHIS. *In* J. L. Sharp & G. J. Hallman, eds. *Quarantine Treatments for Pests of Food Plants*, pp. 1-10. Westview Press, Boulder, Colorado.

Sharp, J. L. 1988. Status of hot water immersion quarantine treatment for Tephritidae immatures in mangos. Proc. Florida State Hort. Soc. 101: 195-197.

Sharp, J. L. 1992. Hot-air quarantine treatment for mango infested with Caribbean fruit fly (Diptera: Tephritidae). J. Econ. Entomol. 85: 2302-2304.

Sharp, J. L. 1993. Hot-air quarantine treatment for 'Marsh' white grapefruit infested with Caribbean fruit fly (Diptera: Tephritidae). J. Econ. Entomol. 86: 462-464.

Sharp, J. L. 1994. Hot water immersion, *In* J. L. Sharp & G. J. Hallman, eds. *Quarantine Treatments for Pests of Food Plants.* pp 133-147. Westview Press, Boulder, Colorado.

Sharp, J. L. & W. P. Gould. 1994. Control of Caribbean fruit fly (Diptera: Tephritidae) in grapefruit by forced hot air and hydrocooling. J. Econ. Entomol. 87: 131-133.

Sharp, J. L. & G. J. Hallman. 1992. Hot air quarantine treatment for carambolas infested with Caribbean fruit fly (Diptera: Tephritidae). J. Econ. Entomol. 85: 168-171.

Sharp, J. L. & R. G. McGuire. 1996. Control of Caribbean fruit fly (Diptera: Tephritidae) in navel oranges by forced hot air. J. Econ. Entomol. 89: 1181-1185.

Shellie, K. C. 1994. Quality of 'Manila' mango after heat treatments for fruit fly disinfestation. HortScience 29: 483.

Shellie, K. C. & R. L. Mangan. 1994. Disinfestation: effect of non-chemical treatments on market quality of fruit. *In* Champ, B. R., E. Highley & G. I. Johnson, eds. *Postharvest Handling of Tropical Fruits*, Proceed. of an Internat. Conf. Held at Chiang Mai, Thailand, 19-23 July 1993. ACIAR Proceed. #50. pp 304-310. Watson Ferguson and Co., Australia. 500 pp.

Shellie, K. C., R. L. Mangan & S. J. Ingle. 1997. Tolerance of grapefruit and Mexican fruit fly larvae to heated controlled atmospheres. Postharvest Biol. and Technol. 10: 179-186.

Shelton, M. D., V. R. Walter, D. Brandl & V. Mendez. 1996. The effects of refrigerated, controlled-atmosphere storage during marine shipment on insect mortality and cut-flower vase life. HortTechnology 6: 247-250.

Soderstrom, E. L. & D. G. Brandl. 1984. Low-oxygen atmosphere for postharvest insect control in bulk-stored raisins. J. Econ. Entomol. 77: 440-445.

Soderstrom, E. L., D. G. Brandl & B. Mackey. 1992. High temperature combined with carbon dioxide enriched or reduced oxygen atmospheres for control of *Trilobium castaneum* (Herbst) (Coleoptera: Curculionidae). J. Stored Product Research 28: 235-238.

Soderstrom, E. L., D. G. Brandl & B. Mackey. 1996a. High temperature alone and combined with controlled atmospheres for control of diapausing codling moth (Lepidoptera: Tortricidae) in walnuts. J. Econ. Entomol. 89: 144-147.

Soderstrom, E. L., D. G. Brandl & B. Mackey. 1996b. High temperature and controlled atmosphere treatment of codling moth (Lepidoptera: Tortricidae) infested walnuts using a gas-tight treatment chamber. J. Econ. Entomol. 89: 712-714.

Soderstrom, E. L., B. Mackey & D. G. Brandl. 1986. Interactive effects of low-oxygen atmospheres, relative humidity, and temperature on mortality of two stored-product moths (Lepidoptera: Pyralidae). J. Econ. Entomol. 79: 1303-1306.

Taschenberg, E. F., F. Lopez & L. F. Steiner. 1974. Responses of maturing larvae of *Anastrepha suspensa* to light and to immersion in water. J. Econ. Entomol. 67: 731-734.

Thomas, D. B. & R. L. Mangan. 1995. Morbidity of the pupal stage of the Mexican and West Indian fruit flies (Diptera: Tephritidae) induced by hot-water immersion in the larval stage. Fla. Entomol. 78: 235-246.

Vail, P. V., J. S. Tebbets, B. E. Mackey & C. E. Curtis. 1993. Quarantine treatments: a biological approach to decision making for selected hosts of codling moth (Lepidoptera: Tortricidae). J. Econ. Entomol. 86: 70-75.

Whiting, D. C., S. P. Foster & J. H. Maindonald. 1991. Effects of oxygen, carbon dioxide, and temperature on the mortality response of *Epiphyas postvittana* (Lepidoptera: Tortricidae). J. Econ. Entomol. 84: 1544-1549.

Whiting, D. C., S. P. Foster, J. van den Huevel & J. H. Maindonald. 1992. Comparative mortality responses of four tortricid (Lepidoptera) species to a low oxygen-controlled atmosphere. J. Econ. Entomol. 85: 2305-2309.

Whiting, D. C., G. M. O'Connor & J. H. Maindonald. 1996. First instar mortalities of three New Zealand leafroller species (Lepidoptera: Tortricidae) exposed to controlled atmosphere treatments. Postharvest Biol. and Tech. 8: 229-236.

Whiting, D. C., G. M. O'Connor, J. van den Huevel & J. H. Maindonald. 1995. Comparative mortalities of six tortricid (Lepidoptera) species to two high-temperature controlled atmospheres and air. J. Econ. Entomol. 88: 1365-1370.

Whiting, D. C. & J. van den Huevel. 1995. Oxygen, carbon dioxide, and temperature effects on mortality responses of diapausing *Tetranychus urticae* (Acari: Tetranychidae). J. Econ. Entomol. 88: 331-336.

Williamson, M. R. & P. M. Winkleman. 1989. Commercial scale heat treatment for disinfestation of papaya. Paper No. 89-6054, Internat. Summer Meeting, American Soc. Agric. Engineers and Canadian Soc. Agric. Engineers, Quebec.

Williamson, M. R. & P. M. Winkleman. 1994. Heat treatment facilities. *In* R. E. Paull & J.W. Armstrong, eds. *Insect Pests and Fresh Horticultural Products: Treatments and Responses.* pp 249-271. CAB International, Wallingford, United Kingdom.

Yocum, G. D. & D. L. Denlinger. 1994. Anoxia blocks thermotolerance and the induction of rapid cold hardening in the flesh fly, *Sarcophaga crassipalpis.* Physiological Entomol. 19: 152-158.

Yokoyama, V. Y. & G. T. Miller. 1996. Response of walnut husk fly (Diptera: Tephritidae) to low temperature, irrigation, and pest-free period for exported stone fruits. J. Econ. Entomol. 89: 1186-1191.

Zee, F. T., M. S. Nishina, H. T. Chan, Jr., & K. A. Nishijima. 1989. Blossom end defects and fruit fly infestation in papayas following hot water quarantine treatment. HortScience 24: 323-325.

Zoltai, P. & P. Swearingen. 1996. Product development considerations for ohmic processing. Food Technology 50: 263-266

9

Cold Storage of Insects for Integrated Pest Management

Roger A. Leopold

This overview of insect cold storage research primarily relates to integrated pest management (IPM) in those programs where insects and mites are to be mass-reared and released to produce some beneficial result as part of a multi-disciplinary pest control strategy. For the purpose of maintaining focus, the utility or the potential benefit of using mass-reared insects and predatory mites in an IPM program will not be dealt with here. Information on this particular aspect can be obtained by consulting Leppla (1984), Tauber et al. (1985), van Lenteren & Woets (1988), and Dietrick (1989).

Mass rearing of insects and mites to control agricultural pests is recorded in ancient Chinese history, and in the United States of America (U. S.) it has been practiced for over 100 years (Ferguson 1990). A recent compilation of suppliers of beneficial organisms in North America lists 132 companies providing over 120 different species, most of which are either insects or predatory mites (Hunter 1994). The business of raising beneficial organisms for pest control had its start mainly as a cottage industry in the U. S. and in some respects it remains at this stage. However, during the April to September growing season one of the larger American companies now produces over 100 million parasitic wasps and 3 million green lacewings a day (Ferguson 1990). Thus, the industry appears to be developing rapidly and will become increasingly more important in the shift to managing agriculture and greenhouse pests with a minimum use of chemical pesticides.

Insects and mites that are mass-produced under factory-like conditions are those arthropods primarily employed in biological control, such as phytophages, parasitoids, and predators, or it may be the target pest insect itself that has been reared and then sexually sterilized for use in the sterile insect technique (SIT). In line with the often complex strategies involved with implementing an IPM program, increased shelf-life with little or no loss of post release effectiveness continues to be sought by those in the business of supplying insects and mites for control programs. The possibility of using cold storage as an aid to mass-rearing was examined over 60 years ago, its implementation coinciding with the onset of reliable mechanical refrigeration (King 1934, Schread & Garman 1934, Hanna 1935). Subsequently, the use of low temperature has proved to be a valuable tool in mass-production of insects and in their delivery to the release site, and considerable research effort has been expended to use cold storage to increase the utility and economy of rearing an effective biocontrol product.

This review covers research that has been conducted using low temperature to store an insect or mite species to improve its availability to consumers, synchronize a desired stage of development for peak release, provide flexibility and efficiency in mass production, and make available standardized stocks for use in long-term ecological, physiological, or genetic research. Also included is the research on the cold storage of the food source where it requires mass-production of a host or prey used to rear the desired beneficial species to be released in a control program.

Parameters of Cold Storage

Short- vs. Long-term Storage

Some explanation is required to acquaint the reader with the arena in which low temperature storage is currently being used and potential future uses for insects and mites. First, the question of what is long-term as opposed to short-term storage is a nebulous one since the requirements for mass-production and subsequent release of beneficial organisms vary with numerous factors such as the immense diversity in life cycles, host or prey selection, and the physiological or developmental status required to be effective in the field. What may be considered by some to be pushing cold storage to the limits, e.g. 10 days storage to accumulate adult wasp parasitoids before release, cannot be compared to cold storage of diapausing eggs of a phytophagous weevil for more than three months. Depending on the species and the situation, a cold storage period of 10 days may be no less cost effective than a much longer period. For this review, with a few exceptions, I have arbitrarily defined short-term storage as less than 1 month and anything over that as long-term storage.

Dormancy

Dormancy is one of the major strategies employed by insects and mites to survive harsh environmental conditions; it can be used in devising cold storage methods to facilitate mass-rearing (Gilkeson 1990, Tauber et al. 1993). Dormancy for the variety of species covered in this review is defined as either diapause or hibernal quiescence. Figure 9.1 shows a schematic representation of the steps involved in mass-rearing of the parasitoid *Trichogramma brassicae* that was developed by a firm in Switzerland for the commercial control of European corn borer. In this scheme, both diapause and quiescence are used to provide flexibility in mass production and release of this insect. The check marks depict points in this rearing scheme where low temperature is used or could be potentially used to maintain dormancy for both the mass-reared factitious host and parasitoid.

Diapause may be an obligative or facultative interruption in the developmental program of the organism. Facultative diapause is elicited by prior programming usually in form of changes in daylength, temperature, and/or food source occurring over a period of time. In some species the induction of diapause may be invoked by signals occurring over two generations. There can be a maternal effect that can either dictate diapause induction completely or merely modify the threshold response in the next generation (Zaslavski 1988). Diapause is to be distinguished from quiescence in that it is mediated via the endocrine system and is usually specific to developmental stage (Beck 1980, Denlinger 1985, Tauber et al. 1986).

Recent studies on aphidiid parasitoids suggest that, in addition to low temperature and/or short daylength, changes in aphid host plant quality, host aphid morph, and aphid life cycle are involved with diapause induction (Brodeur & McNeil 1989, Polgár et al. 1991, Polgár et al. 1995). Larvae of *Praon volucre* and *Aphidius matricariae* enter diapause without responding to environmental signals or maternal effects when they have developed in oviparae of *Aphis fabae, Myzus varians,* and *M. persicae* (Polgár et al. 1991). Further, diapause was not induced when these two parasitoids were reared in anholocyclic strains of aphids under short days and a temperature of 15°C. However, a cold shock of 10°C could induce both hibernal quiescence and diapause in *P. volucre* in a holocyclic aphid host but only hibernal quiescence in anholocyclic hosts (Polgár et al. 1995). It was suggested that the hormonal titer of the various hosts provides the cues for the developing parasitoid larvae as to whether diapause is to be initiated. This plasticity of dormancy induction, if found to be common among aphid parasitoids, has far-reaching implications for those concerned with their mass-rearing and cold storage.

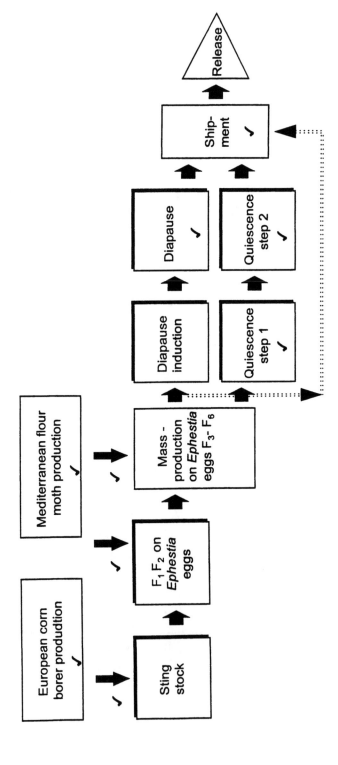

FIGURE 9.1 Schematic illustration of the rearing program for *Trichogramma brassicae* used commercially by a firm in Switzerland. Check marks denote various steps in this program where cold storage can be used to facilitate holding the parasitoid or its mass-rearing host (modified from Bigler 1994)

Thus, not only must the life cycle of the parasitoid be closely monitored but it must be synchronized with the life cycle and availability of suitable hosts on which they are to be reared.

The hibernal quiescence type of dormancy is usually expressed in an organism by a lowering of the metabolic rate, in some cases to an almost undetectable level, and can be elicited by temperature extremes, dehydration, high salt concentrations, and anoxia (Sømme 1995). Hibernal quiescence is more of an immediate response to adverse conditions than is displayed by an organism that is destined to enter diapause. Once the adverse conditions are removed, the organism can quickly recover from quiescence and quickly regains its ability to carry on its life cycle. In contrast, the diapause state ordinarily requires additional signals to trigger recovery such as change in daylength, temperature, and/or moisture, and following this there is usually a lag time during which certain physiological processes again initiate the developmental program. It should be pointed out that all insects and mites do not have the capability to enter diapause nor the ability to maintain hibernal quiescence. In addition, these states of dormancy cannot be maintained indefinitely. There is a metabolic cost for maintaining dormancy and also, if the external conditions are too severe or last too long, damage to the organism does occur. For example, insects and other terrestrial arthropods may enter a chill coma when exposed to temperatures between 0 and 10°C which is ordinarily reversible providing the exposure is not too long (Lee 1991).

Cold Hardening

The term cold hardening refers to physical and/or metabolic modifications occurring within an organism that enable it to survive low temperatures common to an overwintering situation in the north temperate region. Development of cold hardiness may or may not be directly linked to diapause. Denlinger (1991) gives an excellent analysis of the relationships between diapause and cold hardening. He delineates four possible situations: (1) cold hardiness in the absence of diapause; (2) cold hardiness occurring coincidentally with diapause, each having different regulatory signals; (3) cold hardiness occurring as an element of diapause sharing regulatory signals and; (4) diapause occurring in the absence of cold hardiness. Whereas diapause is an all-or-none response, cold hardening is expressed as cold tolerance that is usually elicited proportional to the external stimulus. Initiation of cold hardening, like diapause, depends on environmental cues, commonly low temperature, but may also include daylength and desiccation. Other terms often used in the literature concerned with cold hardening process are cold conditioning, acclimation and acclimatization.

Lee (1989, 1991) has extensively reviewed the various mechanisms which terrestrial arthropods use to survive short-term and long-term exposure to low temperatures that would ordinarily be lethal to these organisms. There are basically two major strategies that terrestrial arthropods employ: freeze tolerance and freeze avoidance. Arthropods which overwinter by avoiding freezing do so by regulating the temperature at which their body water spontaneously freezes. They have the capacity to lower the inherent crystallization point of their body fluids to a temperature not likely to be encountered in an overwintering situation. Cold hardening for these organisms typically involves elimination of heterogenous nucleators and accumulation of anti-freeze proteins or low-molecular-weight polyols and sugars. Freeze tolerant species are able to decrease their capacity to supercool and initiate extracellular ice nucleation which presumably allows for a slowed dehydration of the cells and lessens the likelihood of osmotic shock occurring during a freezing episode (Zachariassen & Hammel 1976, Duman & Horwath 1983). While cold hardening for these organisms may also involve accumulation of low-molecular-weight cryoprotectants, some insects have been found to produce ice-nucleating proteins and lipoproteins (Duman et al. 1985).

The message to be remembered by those attempting to use diapause, quiescence, and/or cold hardening to extend or facilitate a cold storage period is that there can be very precise regulatory cues that involve preconditioning an insect or mite species for storage and, in the case of diapausing species, post conditioning prior to release. These regulatory factors must be incorporated into the mass-rearing regime before cold storage. Failure to pay strict attention to these factors will result in an inferior product for release. For example, the parasitoid *Bracon cephi* is capable of entering diapause and also becoming cold hardened, an apparent coincidental relationship (Salt 1959). However, the cold hardening process lags significantly behind induction of diapause. Thus, placing this insect into a cold storage situation before it has become sufficiently cold hardened could produce irreparable damage.

Chilling, Freezing, and Vitrification

Currently, cold storage for most species that are used in biocontrol, SIT, or an IPM program involves chilling at some subambient temperature above $0°C$. It is not clear whether this is indicative of the low temperature tolerance profile of this particular group of organisms or some other reason. It is a fact that many insects and mites are able to overwinter at sub-zero temperatures in a supercooled state (Lee 1991), yet very few studies have been conducted thus far which examine storage temperatures below 0°C to facilitate mass production and/or release. A possible explanation for the lack of studies examining sub-zero storage is that a major portion of prior research accomplished in this area was concerned with only short-term storage (Table 9.1). The amount of time required to do the necessary cold conditioning of an insect or mite, assuming it had the capacity to tolerate sub-zero temperatures, would often defeat the purpose for short-term storage. In most cases, short-term storage has been used in the range of 3-15°C to accumulate organisms for shipment to a consumer, to synchronize a specific stage for peak release or to maintain them in a quiescent state during shipment to prevent damage or loss. Long-term storage would benefit a program and be cost effective, in terms of maintaining the low temperatures and conducting the necessary pre-programming treatments, where it was desired to shut down mass-rearing during the off season, to maintain a inventory of less used or back-up strains and to preserve genetically important stocks for research.

There was no evidence encountered in this review that any of the various insects or mites having the potential to be used in IPM programs had a freeze tolerant stage in their development. It may be that freeze tolerance has yet to be examined for many of these species or that freeze tolerance is not common among this group. However, freezing of certain developmental stages of host and prey insects has been proven to be a beneficial practice in the mass-rearing of predators and parasitoids for control programs. This is discussed elsewhere in the section dealing with refrigerated and liquid nitrogen storage of host material for mass rearing.

The process of vitrification has probably a limited application in the cold storage of insects and mites for use in IPM programs. Vitrification of a liquid to a solid "glassy" amorphous phase, without formation of ice crystals, occurs only upon ultra rapid cooling to liquid nitrogen temperature for storage (-196°C). Because of their chilling intolerance, certain dipteran embryos are unable to be frozen and stored by conventional cryopreservation methods, which include a slow cooling step (Heacox et al. 1985, Mazur et al. 1992b, Miles & Bale 1995). Preservation of embryos of the pomace fly, *Drosophila melanogaster*, with a high rate of survival, has been accomplished using the vitrification process, a multi-molar cryoprotectant, and liquid nitrogen storage at -196° C (Mazur et al. 1992a, Steponkus & Caldwell 1993). Limited success has also been obtained using a vitrification technique with embryos of

TABLE 9.1 Examples of various insect and acarine taxa examined for cold storage potential in biological control and sterile insect technique (SIT)

Order Family Species	Biocontrol/ SIT Usage	Stage Stored	Storage Temp. (°C)	Length of Storage Period	Reduction in Emergence (E), Lifespan (L), and/or Reproduction (R)	Reference
Heteroptera						
Anthocoridae						
Orius laevigatus	Predator	Adult	13/3	50 days	Yes (L)	Rudolf et al. (1993)
Pentatomidae						
Podisus maculiventris	Predator	Egg	9	6 days	No	de Clerq & Degheele (1993)
		Adult	9	2 months	Yes (L, R)	
Coleoptera						
Coccinellidae						
Leis axyridis	Predator	Adult	5	100-140 days	No	Deng (1982)
Hippodamia convergens	Predator	Adult[1]	3	4-6 months	No	Davis & Kirkland (1982)
Curculionidae						
Trichosirocalus horridus	Phytophage	Egg[1]	4	120 days	No	Kok & McAvoy (1983)
Staphylinidae						
Aleochara bilineata	Parasitoid	Adult[1]	--	2-3 months	No	Adashkevich & Perekrest (1977)
Diptera						
Calliphoridae						
Cochliomyia hominivorax	SIT	Puparium	10	1-3 days	No	Linquist et al. (1992)
Chrysomya bezziana	SIT	Puparium	10	1-7 days	Yes (E, L, R)	Spradbery (1990)
		Adult	8-10	1-10 days	Yes (L)	Spradbery (1990)

Cecidomyiidae						
Aphidoletes aphidimyza	Predator	Larva[1]	5[2]	8 months	No	Gilkeson (1990)
Tachinidae						
Eucelatoria bryani	Parasitoid	Puparium	5	18 days	No (E)	Shekharappa et al. (1988)
Ceranthia samarensis	Parasitoid	Puparium[1]	<8	8 months	--	Mills & Nealis (1992)
Tephritidae						
Bactrocera oleae	SIT	Pupa, Adult	5-11	2 months	Yes (E, L, R)	Tzanakakis & Stamopoulos (1978)
Neuroptera						
Chrysopidae						
Chrysoperla carnea	Predator	Adult[1]	5[2]	31 weeks	No	Tauber et al. (1993)
Chrysopa perla	Predator	Cocoon	6	1-12 months	Yes (E, R)	Canard (1971)
Hymenoptera						
Aphidiidae						
Aphidius matricariae	Parasitoid	Mummy[3]	8[2]	7-30 days	No (L)	Shalaby & Rabasse (1979)
Aphidius smithi	Parasitoid	Mummy,	1-4	>2 months	--	Stary (1970)
		Adult	10	10 days	No	Stary (1970)
Ephedrus cerasicola	Parasitoid	Mummy	0	1-6 weeks	No (L, R)	Hofsvang & Hågvar (1977)
Aphelinidae						
Aphelinus asychis	Parasitoid	Mummy	4.4	15-120 days	Yes (L, R)	Archer & Eikenbary (1973, 1976)
						Whitaker-Deerberg et al. (1994)
Aphtis spp.	Parasitoid	Pupa	15	3-6 days	Yes (E)	Javier et al. (1991)
Encarsia formosa	Parasitoid	Mummy	3-12	3-35 days	Yes (E, L, R)	Liu & Tian (1987)
		Adult	12	5 days	Yes (R)	Liu & Tian (1987)

(continues)

TABLE 9.1 (continued)

Order Family Species	Biocontrol/ SIT Usage	Stage Stored	Storage Temp. (°C)	Length of Storage Period	Reduction in Emergence (E), Lifespan (L), and/or Reproduction (R)	Reference
Braconidae						
Bracon spp.	Parasitoid	Adult	2-10	1-4.5 months	Yes (L, R)	Kovalenkov & Kozlova (1981) Ahmed et al. (1982) Temerak (1984) Jayanth & Nagarkatti (1985)
Encyrtidae						
Leptomastix dactylopii	Parasitoid	Mummy, Adult	5-15	5-40 days	Yes (E, L)	Krishnamoorthy (1989)
Holcothorax testaceipes	Parasitoid	Mummy[1]	-5-5	5-35 weeks	Yes (E)	Wang & Liang (1989)
Eupelmidae						
Anastatus sp.	Parasitoid	Larva	10-12	6 months	No	Huang et al. (1974)
Ichneumonidae						
Campoletis chlorideae	Parasitoid	Pupa	8	10-35 days	Yes (E)	Patel et al. (1988)
Pteromalidae						
Pteromalus puparum	Parasitoid	Prepupa	10	15 months	Yes (E)	McDonald & Kok (1990)
Muscidifurax raptor	Parasitoid	–[4]	6-10	30 days	No (E)	Fabritius (1980)
Scelionidae						
Telenomus remus	Parasitoid	Larva[5]	10	6-16 days	Yes (E)	Gautam (1987b)
Telenomus theophile	Parasitoid	Larva[5]	4	>2 months	No	Liang & Hu (1989)

Trichogrammatidae						
Trichogramma spp.	Parasitoid	Larva[5]	8-12[2]	1-4 months	No (E, R)	Iacob & Iacob (1972) Vigil (1971) Gou (1985) Zhu & Zhang (1987) Ma et al. (1988) Shi et al. (1993) Sun & Yu (1988)
Acarina						
Phytoseiidae						
Amblyseius cucumeris	Predator	Larva, Adult	-8-9	10 weeks	Yes (L)	Gillespie & Ramey (1988)
Phytoseiulus persimilis	Predator	Larva, Adult	7.5	4 weeks	No	Morewood (1992)

[1]Diapausing
[2]Cold conditioned prior to storage
[3]Immature stage stored within mummified insect host
[4]Stored within host puparium
[5]Stored with host eggs

three other dipteran species (Leopold 1998). It is envisioned that inventories of less often used strains, founder strains, geographical isolates, and genetically important stocks could be maintained as vitrified embryos. The benefit of developing such techniques for storage compared with conventional dormancy/cold hardening, if even possible for a desired species, is that preservation at liquid nitrogen temperatures is theoretically perpetual with little chance of genetic or somatic deterioration.

Cold Storage for Mass Rearing and Release

There is tremendous variation in the capacity of insect and mite species to survive a cold storage period as is illustrated in Table 9.1. For this reason, taxonomic generalizations related to cold storage tolerance on the basis of orders, families, and sometimes genera are almost impossible to make. It should be mentioned that a number of the studies cited in Table 9.1 list reductions in emergence, lifespan, and/or reproduction following a cold storage period. However, the magnitude of these detrimental effects caused by the low temperature storage is generally proportional to the length of time spent in storage. Thus, those persons managing mass production of insects or mites have the option to maintain the particular species in cold storage for as long as it meets all the requirements for a certain program. Further, in studies where a series of storage temperatures were tested, there was ordinarily an optimum storage temperature identified where little damage was realized over a portion of the full length of the storage period that was chosen to be examined. For example, 7 days storage at 10°C for the hymenopteran parasitoid *Telenomus remus* within host eggs of *Spodoptera litura* was found to be optimum with no loss of effectiveness of the emerging adults. However, storage at 5 or 15°C caused significant reductions in emergence as did leaving in storage 16 days (Gautam 1986).

Short-term Storage Within Hosts

A common condition for short-term storage of aphid parasitoids is during an inactive stage spent as a larva, prepupa, or pupa within the mummified host. This condition of aphid mummification affords added benefits for insectary management because it provides the parasitoids with a protective encasement that allows easy collection and handling during mass-rearing. Generally, cold storage was most successful when the parasitoids were placed at low temperatures shortly after mummification of the host. Three-day-old mummies containing the parasitoid *Aphelinus asychis* held at 4.4°C yielded the best adult emergence after 3 days of storage. Longer storage periods and older mummies reduced survival (Whitaker-Deerberg et al. 1994). Storage of another aphelinid parasitoid, *Encarsia formosa*, within the mummified aleyrodid host at temperatures ranging from 3-12°C showed that the rate of adult emergence was positively correlated with storage temperature and negatively correlated with the length of storage. Storage at <12°C for 20 days for this species yielded an emergence of adults similar to that of unstored insects (Liu & Tian 1987). The latter study showed a higher emergence rate when mummies were stored on the leaves as opposed to removing them, suggesting that reduced handling or perhaps increased humidity during storage aided survival.

Survival was also highest when *Lysiphlebus testaceipes* was stored in 3 day-old, as opposed to 6 day-old, mummies of *Schizaphis graminum* at temperatures of either 4.4 or 7.2°C (Archer et al. 1973). Storage at 10°C allowed many adults to emerge and even a few at 7.2°C, which indicates that 7°C is near the threshold for adult development for this species. This is in contrast to the encyrtid parasitoid *Leptomastix dactylopii* which did not survive storage in mummies at <15°C (Krishnamoorthy 1989).

Short-term cold storage tolerance during the developmental cycle of *Muscidifurax raptor*, a parasitoid of the house fly, was examined by Fabritius (1980). Parasitoid losses were at a minimum when the host house fly puparia were put into storage 11-13 days after parasitization at 6-10°C for up to 30 days, indicating that cold tolerance varies with developmental stage even though a major portion of the parasitoid's life cycle is spent within the host.

Parasitized host eggs provide another possible stage for short-term storage of the hymenopteran egg parasitoids. *Trichogramma* spp. having an 8 day developmental cycle before emerging from *Ephestia kuehniellla* eggs, could tolerate 30 days storage within eggs at 10°C without adverse effects (Vigil 1971). As with other insects, it is essential to determine the developmental stage having the most cold tolerance to gain peak emergence and efficacy of the mass-reared parasitoids following storage. The optimum age for storage of *S. litura* eggs parasitized by *T. remus* was 7 days after the initial parasitization (Gautam 1986).

Short-term Storage With No Host

There are situations where it is desirable to hold adult parasitoids or various developmental stages of non-parasitoids under short-term cold storage. For instance, the egg and adult stage of predatory pentatomids, *Podisus* spp., can be stored at 9°C for 6 and 30 days, respectively, without adverse affects on survival, lifespan, or reproduction (de Clercq & Degheele 1993). Embryos of an age of less than 1 day post-oviposition were reported to be extremely chilling intolerant and there was an apparent species difference in cold tolerance for the adult stage. *P. maculiventris* could be stored for up to 2 months without negative effects while the limit for *P. sagitta* was 1 month. Shi et al. (1993) also observed a difference in cold storage tolerance in several geographical strains of *Trichogramma dendrolimi;* those coming from northern China tolerated storage better than strains isolated from the southern regions.

Accumulation, shipment, and release of sexually sterile insects as part of a SIT or IPM program often present mammoth logistical problems. A 1991 eradication program of the New World screwworm, *Cochliomyia hominivorax*, in Libya using flies mass-reared in Mexico, demanded that emergence of the sterile pupae be synchronized using holding temperatures of 10, 20, and 26°C in order that four batches of 10 million flies per week could be released over a period of about six months (Lindquist et al. 1992). Cold storage of the Old World screwworm puparia is also essential to the mass-rearing operations. Five-day-old puparia of *Chrysomya bezziana* are stored at 8-10°C overnight to maintain a twice weekly egging schedule for the factory colony flies (Spradbery 1990). Also, the irradiated puparia of this insect can be accumulated prior to shipment and release at 9-10°C for up to 3 days without loss of emerging adults. Interrupting the quiescent state by giving the puparia of *Musca domestica, Phaenicia sericata,* and *Lucilia cuprina* intermittent recovery periods at 25-28°C enhances the ability of these flies to survive storage at 10°C (Leopold et al. 1998).

Short-term Acclimation for Short-term Storage

As mentioned above, extensive preprogramming of an insect or mite species to enable it to cope with low temperature ordinarily defeats the purpose for short-term storage. However, if the preprogramming period is not lengthy or complicated, it may be profitable to conduct a pre- or post storage manipulation of rearing conditions. Preacclimation of *Aphidius matricariae* and *Trioxys indicus* within host mummies at temperatures of 12-15°C for up to 72 h before placing into storage at 8 and 4°C, respectively, increased emergence (Shalaby & Rabasse 1979, Singh & Srivastava 1988). Emergence for *A. matricariae* was near 90% after 15 days storage and about 72% for *T. indicus* after 30 days. A deacclimation regime consisting of raising the temperature 3°C every 12 h up to ambient from a storage temperature of 4.4°C was used in the recovery of stored *A. asychis* adults (Archer et al. 1976). With this method, adults stored up to 15 days were as effective as unstored but reproduction declined as storage time was increased.

Cycling the storage temperature between 3 and 13°C was found to be better than maintaining at 9°C for two anthocorid predators. Fecundity was not reduced for *Orius laevigatus* after 50 days in storage, although that of *O. majusculus* was substantially reduced after 20 days (Rudolf et al. 1993).

Long-term Storage

Many of the reports concerning the long-term storage of insect or mite species to be used in biocontrol programs involve some degree of preprogramming for dormancy and/or cold conditioning. Placing the mass-reared colony females of the curculionid *Trichosirocalus horridus* under short days and a 21/10°C alternating day/night temperature induced diapause in their progeny. Eggs of this phytophagous biocontrol agent for thistle could be kept at 4°C for up to 120 days with no loss in emergence or viability (Kok & McAvoy 1983). Further, the post storage emergence and fecundity of four species of *Trichogramma* could be enhanced by alternating the day-night temperature for the parasitized host eggs between 23-25°C and 8-12°C (Gou 1985, Zhu & Zhang, 1987). With this type of regime, *T. ostriniae* had an 88% emergence rate and produced 55 eggs/female after 90 days storage. Three to four day-old pupae of *Bracon hebetor* could be stored for 60 days under a temperature regime alternating between 2-5 and 30°C (Ahmed et al. 1982). Interestingly, rearing and storage of this wasp parasitoid was being investigated so it could be used as a means to reduce the population of *Ephestia cautella* to a level where it would be feasible to initiate a SIT program for this pest. In these latter three studies, diapause induction associated with the temperature cycling was not investigated for these species.

Gilkeson (1990) showed that storage of *Aphidoletes aphidimyza* cocoons at ≤ 11°C in total darkness was a sufficient external signal to induce diapause even though the larvae were reared under non-diapausing conditions. Midge larvae stored up to 8 months at 5°C had <10% mortality, and the resulting adult females oviposited a normal complement of eggs. In another study, duration and temperature of storage were found to be interacting factors in the termination of diapause in pupae of the encyrtid parasitoid *Holcothorax testaceipes.* Longer storage periods (ca. 15 wk) and a higher storage temperature (5 vs. 0 and -5°C) were associated with shorter times of morphogenesis and duration of emergence, while lengthening the photophase had no effect (Wang & Liang 1989). Similar results were found for breaking diapause of the ichneumonid *Exetastes cinctipes.* The proportion of individuals breaking prepupal diapause for this insect depended upon the nature of the rearing conditions during the embryonic and larval stages in addition to storage temperature and duration during the pre-pupal stage (Slovak 1988).

Diapausing adults of the predator *Chrysoperla carnea* were stored over 6 months under short daylengths at 5°C and retained high post storage survival and fecundity (Tauber et al. 1993). In this case, diapause induction was during the larval stage, and then the emerging adults were submitted to a 5 week cold acclimation regime during which the temperature was decreased stepwise from 21 to 10°C prior to placing in long-term storage. Storage of *Chrysopa perla* in the cocoon stage was accomplished in the absence of diapause induction and acclimation for 6 months at 6°C. The fecundity of stored females was equal to that of females emerging from diapause and the emergence was greater than half the normal laboratory rate (Canard 1971).

Feeding and Cold Storage

Providing a source of nutrition and water during cold storage was usually found to be beneficial when the feeding stages of insects and mites were held at temperatures where they remained active. The storage temperatures at which feeding activity was apparently maintained varies with the individual species. For example, providing food and a moist environment was found by Morewood (1992) to be preferable for storage of the mite *Phytoseiulus persimilis* even though the storage temperature of 8°C was lower than the threshold for development. Providing food and water for 10 to 30 days before storage clearly enhanced post storage survival of the coccinellid *Leis axyridis* (Deng 1982), and it was also considered necessary for *C. carnea* during the 5 week prestorage acclimation period as well as during storage at 5°C (Tauber et al. 1993). Stary (1970) reported that supplying a honey-water mixture prior and during shipment was found to improve post release survival of newly emerged *A. smithi* parasitoids. An alternative strategy of removing *A. asychis* adults from storage weekly for a feeding period was reported by Archer et al. (1976) to enhance survival. It was found to be superior to monthly feeding schedule, while providing a honey solution at 28°C once every 2 wk was sufficient nutrition during storage for *Bracon* sp. (Kovalenkov & Kozlova 1981). A more rigorous schedule of removing from storage and feeding olive fly adults 3 times a week was found to be essential for their survival and maintenance of fecundity (Tzanakakis & Stamopoulos 1978).

Providing water and food before or during storage may not be advisable in all situations, especially if it is desired to emulate a known sub-zero overwintering condition for a particular species. Ingestion of food and even just water has been correlated with loss of cold hardiness (Sømme 1966, Baust & Morrissey 1975, Young & Block 1980, Cannon et al. 1985). Heterogenous gut contents have been suggested to serve as incidental ice nucleators causing lethal freezing in freeze intolerant species (Salt 1936, 1953, Zachariassen 1985). Hofsvang & Hågvar (1977) noted that the digestive tract of *E. cerasicola* emptied about the second to third day after mummification of its host, and they suggested that this may be one of the factors which enable it to tolerate sub-zero temperatures. Further, in the coccinellid *Coleomegilla maculata* Baust & Morrissey (1975) showed that cold hardiness diminished significantly when only distilled water was provided. The supercooling point of another lady beetle, *Hippodamia convergens*, was shown to increase markedly from -16 to near -3°C after ingestion of ice nucleating bacteria that commonly grow epiphytically on the surface of plants (Strong-Gunderson et al. 1990). Thus, in cases where subzero storage of a feeding stage is contemplated, it is necessary to know whether cessation of feeding, seasonal starvation or sufficient time for a species to clear its gut contents is part of its normal preparation for surviving exposure to sub-zero temperatures.

Variation in Gender Response to Cold Storage

Numerous observations have been made on the variation in ability of the sexes to survive a storage situation. These variations can have a bearing on the outcome of control programs. For example, distortions in survival rate between the sexes would not be as important in the storage of parasitoids, providing it favored the females, as it would be for predators or phytophages. With the hymenopteran parasitoids, survival during and following cold storage generally favors the females regardless of the stage of development that was stored (Archer & Eikenbary 1973, Hofsvang & Hågvar 1977, Kovalenkov & Kozlova 1981, Jackson 1986, Zhu & Zhang 1987, Zhang 1992, Whitaker-

Deerberg et al. 1994). Exceptions to these observations were those of Krishnamoorthy (1989) and Patel et al. (1988) on *Leptomastix dactylopii* and *Campoletis chlorideae* who found the females to be less cold tolerant than males under their conditions for storage. Further, Shalaby & Rabasse (1979) found no difference in post storage adult lifespan between the sexes of *A. matricariae* after short-term storage as either adults or diapausing larvae within mummies. More bewildering is the report of Jayanth & Nagarkatti (1985) who found survival throughout storage for *B. brevicornis* to favor the females but as duration of cold storage increased, so did the proportion of surviving males.

With neuropteran predators, *C. perla* females survived storage better than males when stored within their cocoons while diapausing adults of *C. carnea* showed no difference in survival between the sexes either within or following cold storage (Canard 1971, Tauber et al. 1993).

With the exception of hymenopteran parasitoid males, reduction in survival of the male species caused by a period of cold storage will have a greater immediate impact on the relative effectiveness of biocontrol or SIT in an IPM program. For example, loss of predacious males results in loss of functional biological control agents while loss of hymenopteran parasitoid males, which typically have facultative arrhenotokous mates, represents a less important factor in an overall control program. Further, it has been shown that post cold storage reproduction of *A. asychis* could occur normally in the absence of surviving males as long as the females mated prior to being placed in storage (Archer & Eikenbary 1973). Cold storage had little effect on the fertilizing ability of sperm within these females as their F_1 progeny ratio of females to males was similar to unstored controls.

There have been reports that indicated some species have benefited by being placed into cold storage. Mature larvae of 3 species of pteromalid parasitoids responded positively to up to 180 days of cold storage. The resulting adult females of *M. raptor* and *M. zararaptor* showed an increase over nonstored females in lifespan, fecundity, and produced progeny with a greater total weight while *Spalangia endius* produced larger progeny and had an increased lifespan (Legner 1976). For two *Trichogramma* spp., fecundity and the progeny female: male ratio was greater than controls after 4 months cold storage (Zhu & Zhang 1987). Neither of these two studies mentioned whether diapause occurred for these species during the storage period. However, in the latter study the insects were treated with a prestorage acclimation regime by alternating the day-night temperature. An interesting outcome of cold storage was observed by Petters & Grosch (1977) for *B.*

hebetor females. Supernumerary ovarioles were produced by placing the fourth-instar larvae into cold storage, thus increasing the reproductive potential of the adult females.

Quality Control Testing After Cold Storage

Besides the obvious indicators that can be assayed to reveal possible reduced post cold storage quality (emergence, lifespan, fecundity, progeny ratios, and reproductive success), there are other factors which impact on post release effectiveness of mass-reared biocontrol agents. Bigler (1994) analyzed the requirements for quality control testing in *Trichogramma* and concluded that a combination of quality traits having high correlation to field effectiveness should be established for each step in the rearing and release program. For example, post release dispersal and host or prey seeking ability are important attributes to be examined in species reared for control purposes. Dispersal or flight ability has been assayed in a several biocontrol species after a period of cold storage. Davis & Kirkland (1982) noted a difference in the propensity for post cold storage dispersal relating to whether the lady beetle *H. convergens* had been fed. Unfed beetles had a greater dispersal rate 3 days following storage than did the fed beetles. Another beetle predator, *Rhizophagus grandis*, lost flight propensity steadily as the cold storage period increased (Coullien & Gregoire 1994). The authors suggested that reduction in adult fat body reserves during cold storage could be a reason for this loss. They also noted that cold storage had no affect on the response of the beetles to a synthetic feeding attractant used in the culture medium for mass-rearing.

Examining the rate of post storage parasitism by parasitoids offered an unlimited number of hosts would appear to be another means of testing fecundity, host attraction, and dispersal ability, providing the test was properly designed. However, few studies have been conducted that cite such data that specifically relate to cold storage. Piao et al. (1992) found no difference in post storage parasitism rate and progeny production for *T. dendrolimi* females stored within parasitized eggs up to 30 days at 3-5° C. In a similar study, Iacob & Iacob (1972) observed a reduction in parasitism rate after storage of *T. evanescens* as eggs or larvae within the host stored at the higher temperature range of 9-12° C. Another difference between the two studies, was that the former employed a period of cold conditioning at 10° C which may have aided in maintaining post storage quality.

After storage at 4-6°C, adult females of the phytophagous chrysomelid *Zygogramma suturalis* displayed earlier ovipositional peaks than did unstored controls. Whether this effect is caused by ovarian development occurring within storage or by a post storage response to the stress of low temperature was not investigated (Wan & Wang 1990).

Comparison of methods for placing sterilized adults of the melon fly, *Bactrocera cucurbitae,* in a quiescent state during shipment prior to release in a SIT program was found to favor chilling as opposed to anoxic anaesthetization with either CO_2 or N_2 (Tanahara & Kirihara 1989). Recovery was quicker regardless of the duration of the chilling period and the flight ability was curtailed more severely by CO_2 exposure than by chilling or N_2.

Cold Storage of Host or Prey to Facilitate Mass-Rearing

The availability of artificial diets for most of the entomophagous species currently used in biological control is limited and thus a host or prey food source must also be mass-reared as part of the production scheme. Long-term, low temperature storage has been used for the eggs from a variety of insects and will support the growth of a number of oophagous and endoparasitoid biocontrol species (Table 9.2). There have been considerably fewer studies conducted on storage of the post embryonic stages, but these reports have also been generally favorable.

Because it is often not necessary for the host or prey species to survive the storage period for use in mass-rearing, the temperature range used in cold storage and the length of storage can be much more extreme than can be currently employed in storage of the parasitoid or predator. Storage of host and prey species has been accomplished at temperatures ranging from 15 to - 196° C and the time frame for storage of eggs of several pentatomid species has been extended up to 5 years with little loss in the usefulness for rearing parasitoids (Table 9.2).

Refrigerated Storage of Host or Prey for Mass-Rearing

The refrigerated storage period of an egg host in a chilled or frozen state for use in mass-rearing of *Trichogramma spp.* is generally useful only on a short-term basis (Voegele et al. 1974, Medina & Cadapan 1982, Pizzol & Voegele 1988, Pu et al. 1988). The longest refrigerated storage of a host was for 4 months for *Samia cynthia ricini* eggs held at - 18°C (Pu et al. 1988). However, in this case, it was reported that continuous rearing of the parasitoids on factitious host eggs reclaimed from frozen storage resulted in low production. Likewise, Voegele et al. (1974) found that *E. kuehniella* eggs could be stored unfrozen up to 2 months at 4°C after which they suggested that an unknown factor was lost that was required for development of *T. brasiliensis* and *T. evanescens* embryos. A novel attempt to extend the useful shelf-life of host eggs was made by Ramos & Jimenez (1993). Eggs of another lepidopteran, *Sitotroga cerealella*, were vacuum packed at pressures up to 10 PSI and stored at 7°C for 20 to 60 days. While the effective storage time for this method did not extend past 35 days, a very high rate of parasitism by *Trichogramma spp.* was observed for the eggs treated in this manner.

There is species variation among the *Trichogramma* in the acceptance of cold-stored host eggs. *T. chilotraeae* showed no preference between cold-stored and fresh eggs of *Corcyra cephalonica*, while *T. australicum* would not parasitize eggs that had been refrigerated (Medina & Cadapan 1982). Further, a temporal decline in the acceptability of cold-stored host eggs has been observed for several trichogrammatid and non-trichogrammatid species. The rate of parasitism for *T. exiguum* fell after host eggs of *C. cephalonica* had been stored at -6°C for longer than 8 days (Dass & Ram 1985). Similar results showing a decrease in host acceptability with storage duration were found for *Telenomus remus* with eggs of *Spodoptera litura* and for *Ooencyrtus ennomophagous* with *Clostera inclusa* eggs (Gautam 1987a, Drooz & Solomon 1984). With *O. ennomophagous*, a situation unrelated to host acceptance was observed when it was offered eggs from a different lepidopteran. Eggs from the geometrid, *Lambdina pellucidaria,* after storage at -10°C, were found suitable for production of *O. ennomophagus,* while no parasitoids emerged from fresh eggs (Drooz 1981). This suggests that cold storage may have inhibited an internal developmental or physiological change in a factitious host egg that was ordinarily detrimental to development of the parasitoid.

Storage of house fly puparia at -21°C over a period of 53 wk did not decrease the suitability for rearing the endoparasitoid *M. raptor* (Klunker 1982). The frozen puparia were found to be the most acceptable for rearing *M. raptor* if reclaimed by using a rapid thaw at 65°C. Storage of lepidopteran larvae or pupae by chilling at 5-15°C provides prey source for those parasitoids and predators that are amenable to rearing on these developmental stages. The predacious bug *Oechalia shellenbergii* readily accepts cold-stored *Helicoverpa puntigera* larvae and was reared in greater numbers if offered dead prey rather than alive (Awan 1983). In addition, a closely related species, *H. armigera,* was reported to be able to be maintained in quiescence for up to a year in the pupal stage at 15°C (Giret & Couilloud 1982).

TABLE 9.2 Examples of host or prey placed in cold storage to facilitate mass-rearing procedures

Host or Prey: **Order** Family Species	Developmental Stage Stored	Storage Temp. (°C)	Maximum Storage Time	Mass-Reared Biocontrol Agent (Pa=Parasitoid, Pr=Predator)	References
Heteroptera					
Coreidae					
Gonocerus acuteangulatus	egg	–196	3 years	Gryon spp. (Pa), Trissolcus spp. (Pa), Ooencyrtus spp. (Pa)	Genduso (1980)
Pentatomidae					
Graphosoma lineatum	egg	–196	5 years	Trissolcus simoni (Pa)	Gennadiev & Khlistovskii (1980)
Nezara viridula	egg	–196	3 years	Gryon spp. (Pa), Trissolcus spp. (Pa), Ooencyrtus spp. (Pa)	Genduso (1980)
Scutelleridae					
Eurygaster austriaca	egg	–196	3 years	Gryon spp. (Pa), Trissolcus spp. (Pa), Ooencyrtus spp. (Pa)	Genduso (1980)
Diptera					
Agromyzidae					
Liriomyza spp.	puparium	7	3-6 months	Chrysocharis oscinidis (Pa)	van der Linden (1990)
Muscidae					
Musca domestica	puparium	–7, –21	1 year	Muscidifurax raptor (Pa)	Klunker (1982)

Lepidoptera
Gelechiidae

Sitotroga cerealella	egg	7[1], –196	2-18 months	Trichogramma spp. (Pa)	Gennadiev et al. (1987), Ramos & Jimenez (1993)
		–196	9 months	Chrysoperla carnea (Pr), Chrysopa septempunctata (Pr)	Beglyarov et al. (1981)
Geometridae					
Lambdina pellucidaria	egg	–10	4 weeks	Ooenocyrtus ennomophagus (Pa)	Drooz (1981)
Lymantriidae					
Acyphas leucomelas	egg	–196	1-9 months	Trichogramma dendrolimi (Pa)	Ma (1988)
Noctuidae					
Helicoverpa armigera	pupa	15	1 year	--	Giret & Couilloud (1982)
H. punctigera	larva	5	--	Oechalia schellenbergii (Pr)	Awan (1983)
Spodoptera litura	egg	10	9 days	Telenomus remus (Pa)	Gautam (1987a, b)
Notodontidae					
Clostera inclusa	egg	–10	8-24 months	Ooenocyrtus ennomophagus (Pa)	Drooz & Solomon (1984)
Pyralidae					
Corcyra cephalonica	egg	–196	8 months	Trichogramma spp. (Pa)	Hu & Xu (1988)
C. cephalonica	egg	–6	8 days	T. exiguum (Pa)	Dass & Ram (1985)
Ephestia kuehniella	egg	1, 3	2 months	T. maidis (Pa)	Pizzol & Voegele (1988)
Saturniidae					
Antheraea pernyi	egg	–196	>2.5 years	Trichogramma spp. (Pa)	Wang et al. (1988)
Samia cynthia ricini	larva	10-12	6 months	Anastatus sp. (Pa)	Huang et al. (1974)
S. cynthia ricini	egg	–18	4 months	Trichogramma spp. (Pa)	Pu et al. (1988)

[1]At a vacuum pressure of 10 psi.

Liquid Nitrogen Storage of
Host or Prey for Mass-Rearing

Observations that egg hosts are accepted by many parasitoids in which development has been arrested by freezing or irradiation or never initiated as with unfertilized eggs (Eidmann 1934, Houseweart et al. 1982) have kindled some significant prospects in terms of long-term storage using liquid nitrogen. The early work of Gennadiev & Khlistovskii (1980) using a cryoprotectant and conventional cryopreservation methods to store *G. lineatum* and *S. cerealella* eggs for later use in mass-rearing of *Trissolcus* and *Trichogramma* has paved the way for the development of simplified long-term storage methods at liquid nitrogen temperature. Essentially, the eggs are quick frozen without cryoprotectant and also reclaimed by fast thaw procedures. Gennadiev et al. (1987) report that the freezing and thawing steps of *S. cerealella* eggs for rearing of *Trichogramma* must be no longer than 2 sec duration. Hu & Xu (1988) subsequently disclosed that with *C. cephalonica* eggs freezing in liquid nitrogen vapor gave better results than freezing in the liquid phase. Both studies confirmed that rate of parasitism following liquid nitrogen storage was highest when newly oviposited eggs as opposed to older eggs were stored by these methods. It is significant that these methods that promote ultra fast freezing and thawing of host eggs mirror the development of methods for the successful storage and recovery of live *Drosophila* embryos. It was only after the discovery that immersion in liquid nitrogen slush gave the required ultra fast cooling rates, which needed to be accompanied by a recovery with rapid warming, that methodology for useful cold storage of this insect became available (Mazur et al. 1992a, Steponkus & Caldwell 1993).

Other studies also report good to excellent production of parasitoids and predators reared on eggs stored in liquid nitrogen (Beglyarov et al. 1981, Genduso 1980, Ma 1988, Morrison 1988, Wang et al. 1988). In the technique for recovery of *S. cerealella* eggs from liquid nitrogen storage, Gennadiev et al. (1987) advocate drying of the eggs after thawing on hydrophilic plates to gain the best post storage parasitism rate for *Trichogramma*. This suggests that lyophilization following a liquid nitrogen or dry ice/acetone quench might be a procedure that would yield similar high parasitism or predation rates. Whether a certain amount of reconstitution of freeze-dried eggs with water or water vapor would be required for maximum parasitoid acceptance could be easily tested.

Future Directions

Low temperature storage is an integral part of the process of mass-rearing insects and mites for use in agriculture and greenhouse pest control programs. It is the practical application of information provided by researchers studying arthropod cryobiology, dormancy, host-prey interactions, and mass-rearing methods. Cold storage allows insectary managers to gain flexibility and enables them to supply a purely biological product on demand. It is quite evident that the effectiveness of any biological agent used for pest control purposes depends primarily on being released at the proper time. Unforeseeable environmental influences such as those impacting on pest migration, population increase, and crop growth amplify the need for precise timing, especially when releases of insects and mites are to be integrated into multi-disciplinary control programs.

For those involved in rearing of insects and mites as a commercial enterprise, the capability to satisfy the ebb and flow demands for their products will, no doubt, continue to predict their profit or loss margin. Further, expansion into new market areas by making available a new exotic biocontrol species requires that its capacity for low temperature tolerance, diapause and quiescence be assessed before the feasibility for mass-production can be judged. For example, *Tetrastichus gallerucae,* an egg parasitoid of the elm leaf beetle, has been released in California numerous times over a span of 60 years in ineffective efforts to have it permanently established without knowing whether it has an overwintering stage, diapause, or alternate hosts (Boivin 1994).

On the horizon there are definite indications that the capability to insert beneficial genetic characteristics into insects will be forthcoming (O'Brochta & Atkinson 1997). This will provide researchers with tremendous tools for genetically improving and customizing beneficial insects in ways that will increase their value to IPM programs. How these capabilities will change the way pest control strategies will be designed in the future, one can only speculate. Heilmann et al (1993) and DeVault et al. (1996) have profiled a number of possible ways that recent and future bio-technological advances could be used to improve bio-control species for use in IPM programs. Enhancement of hardiness, flight dispersal, pesticide and pathogen resistance, and host-seeking ability were a few of the characteristics that were highlighted for improvement through genetic engineering. Thus, past cold storage successes, the emergence of new bio-control technology, the discovery of promising exotic parasitoids, predators, and phytophages, and the necessity to continue to improve mass-rearing procedures all indicate that a sustained effort should be exerted in developing low temperature storage methods to facilitate the most efficient use of insects and mites in IPM programs.

References

Adashkevich, B. P. & O. N. Perekrest. 1977. Rearing and calculation of effectiveness of *Aleochara*. Zashchita Rastenii 6: 29-30.

Ahmed, M. S. H., A. M. Al Saqur & Z. S. Al Hakkak. 1982. Effect of different temperatures on some biological activities of the parasitic wasp, *Bracon hebetor* (Say) (Hymenoptera). Date Palm J. 1: 239-247.

Archer, T. L. & R. D. Eikenbary. 1973. Storage of *Aphelinus asychis*, a parasite of the greenbug. Environ. Entomol. 2: 489-490.

Archer, T. L., R. K. Bogart & R. D. Eikenbary. 1976. The influence of cold storage on the survival and reproduction by *Aphelinus asychis* adults. Environ. Entomol. 5: 623-625.

Archer, T. L., C. L. Murray, R. D. Eikenbary, K. J. Starks & R. D. Morrison. 1973. Cold storage of *Lysiphlebus testaceipes* mummies. Environ. Entomol. 2: 1104-1108.

Awan, M. S. 1983. A convenient recipe for rearing a predacious bug, *Oechalia schellenbergii* Guerin-Menville (Hemiptera: Pentatomidae). Pakistan J. Zool. 15: 217-218.

Baust, J. G. & R. E. Morrissey. 1975. Supercooling phenomenon and water content independence in the overwintering beetle, *Colemegilla maculata*. J. Insect Physiol. 21: 1751-1754.

Beck, S. D. 1980. *Insect Photoperiodism*. 2nd ed. New York: Academic Press.

Beglyarov, G. A., V. G. Gennadiev, I. A. Ponomareva & E. D. Khlistovskii. 1981. Cryopreserved eggs of the grain moth. Zashchita Rastenii No. 5, 31.

Bigler, F. 1994. Quality control in *Trichogramma* production. *In* E. Wajnberg & S. A. Hassan eds. *Biological Control with Egg Parasitoids*. pp 91-111. Wallingford: CAB International.

Boivin, G. 1994. Overwintering strategies of egg parasitoids. *In* E. Wajnberg & S. A. Hassan, eds. *Biological Control with Egg Parasitoids*. pp 219-244. Wallingford: CAB International.

Brodeur, J. & J. N. McNeil. 1989. Biotic and abiotic factors involved in diapause induction of the parasitoid, *Aphidius nigripes* (Hymenoptera: Aphididiae). J. Insect Physiol. 35: 969-974.

Canard, M. 1971. The possibilities of long-term preservation of the cocoons of a predator of aphids; *Chrysopa perla* L. (Neuoptera, Chrysopidae). Annales de Zoologie, Ecologie Animale 3: 373-377.

Cannon, R. J., W. Block & G. D. Collett. 1985. Loss of supercooling ability in *Cryptopygus antarcticus* (Collembola: Isotomidae) associated with water uptake. Cryo-Letters 6: 73-80.

Couillien, D. & J. C. Gregoire. 1994. Take-off capacity as a criterion for quality control in mass-produced predators, *Rhizophagus grandis* (Coleoptera: Rhizophagidae) for the biocontrol of bark beetles, *Dendroctonus micans* (Coleoptera: Scolytidae). Entomophaga 39: 385-395.

Dass, R. & A. Ram. 1985. Effect of frozen eggs of *Corcyra cephalonica* Stainton (Pyralidae: Lepidoptera) on parasitism by *Trichogramma exiguum* (Pinto and Platner) (Trichogammatidae: Hymenoptera). Indian J. Entomol. 45: 345-347.

Davis, J. R. & R. L. Kirkland. 1982. Physiological and environmental factors related to the dispersal flight of the convergent lady beetle, *Hippodamia convergens* (Guerin-Meneville). J. Kans. Entomol. Soc. 55: 187-196.

de Clercq, P. & D. Degheele. 1993. Cold storage of the predatory bugs *Podisus maculiventris* (Say) and *Podisus sagitta* (Fabricius) (Heteroptera: Pentatomidae). Parasitica 49: 27-41.

Deng, D. A. 1982. Experiments on feeding with artificial diets and cold storage of *Leis axyridis* Pallas. Insect Knowledge Kunchong Zhishi 19: 11-12.

Denlinger, D. L. 1985. Hormonal control of diapause. *In* G. A. Kerkut & L. I. Gilbert, eds. *Comprehensive Insect Physiology, Biochemistry and Pharmacology*. vol. 8: 353-412. Oxford: Pergammon Press.

Denlinger, D. L. 1991. Relationship betweeen cold hardiness and diapause, *In* R. E. Lee & D. L. Denlinger, eds. *Insects at Low Temperature*, pp 174-198. New York: Chapman & Hall.

DeVault, J., K. J. Hughes, O. A. Johnson & S. K. Narang. 1996. Biotechnology and new integrated pest management approaches. Bio/ Tech. 14: 46-49.

Dietrick, J. E. 1989. Commercialization of biological control in the United States. *International Symposium on Biological Control Implementation*. North Amer. Plant Prot. Organ. Bull. 6: 71-87.

Drooz, A. T. 1981. Subfreezing eggs of *Lambdina pellucidaria* (Lepidoptera: Geometridae) alters status as factitious host for *Ooenocytrus ennomophagus* (Hymenoptera: Encyrtidae). Can. Entomol. 113: 775-776.

Drooz, A. T. & J. D. Solomon. 1984. Temporal cold storage of eggs of the popular tent maker, *Clostera inclusa* prior to use in rearing the egg parasite, *Ooencyrtus ennomophagus*. Res. Note, USDA Southern Forest Experiment Station, No. SO 304, 2pp.

Duman, J. G. & K. L. Horwath. 1983. The role of hemolymph proteins in the cold tolerance of insects. Annu. Rev. Physiol. 45: 361-270.

Duman, J. G., L. G. Neven, J. M. Beals, K. R. Olson & F. J. Castellino. 1985. Freeze tolerance adaptations, including haemolymph protein and lipoprotein nucleators, in the larvae of the cranefly *Tipula trivittata*. J. Insect Physiol. 31: 1-8.

Eidmann, H. 1934. Zur Kenntnis der Eiparasiten der Forleule insbesondere über die Entwicklung und Ökologie von *Trichogramma minutum* Riley. Mittellungen Fortwirshaft und Forstwissenshaft 5: 56-77.

Fabritius, K. 1980. Storage of pupae of *Musca domestica* L. parasitised by *Muscidifurax raptor* Gir. & Sand. (Hymenoptera: Chalcidoidea). Studii si Cercetari de Biologie, Animale 32: 83-88.

Ferguson, J. 1990. Better good bugs. Ag Consultant 46: 3-4.

Gautam, R. D. 1986. Effect of cold storage on the adult parasitoid *Telenomus remus* Nixon (Scelionidae: Hymenoptera) and the parasitised eggs of *Spodoptera litura* (Fabr.) (Noctuidae: Lepidoptera). J. Entomol. Res. 10: 125-131.

Gautam, R. D. 1987a. Cold storage of eggs of host, *Spodoptera litura* (Fabr.) and its affects on parasitism by *Telenomus remus* Nixon (Scelionidae: Hymenoptera). J. Entomol. Res. 11: 161-165.

Gautam, R. D. 1987b. Limitations in mass-multiplication of scelionid, *Telenomus remus* Nixon, potential egg-parasitoid of *Spodoptera litura* (Fabricius). J. Entomol. Res. 11: 6-9.

Genduso, P. 1980. Cold storage of eggs of Heteroptera in liquid nitrogen for the rearing of egg parasites. Atti XI Congresso Nazionale Italiano di Entomologia. 365-370.

Gennadiev, V. G., & E. D. Khlistovskii. 1980. Long-term cold storage of host eggs and reproduction in them of egg parasites of insect pests. Z. Obshchei Biol. 41: 314-319.

Gennadiev, V. G., D. E. Khlistovskii & L. A. Popov. 1987. Cryogenic storage of host eggs. Zashchita Rastenii No. 5, 36-37.

Gilkeson, L. A. 1990. Cold storage of the predatory midge *Aphidoletes aphidimyza* (Diptera: Cecidomyiidae). J. Econ. Entomol. 83: 965-970.

Gillespie, D. R. & C. A. Ramey. 1988. Life history and cold storage of *Amblyseius cucumeris* (Acarina: Phytoseiidae). J. Entomol. Soc. B.C. No. 85, 71-76.

Giret, M. & R. Couilloud. 1982. Effect of temperature on the nymphal stage of *Heliothis armigera* Hubn. (Lepidoptera: Noctuidae): technique of storage by arrest of development at 15 deg C. Coton et Fibres Tropicales 37: 271-276.

Gou, X. Q. 1985. Cold storage test of rice moth eggs parasitized by *Trichogramma ostriniae*. Chin. J. Biol. Cont. 1: 20-21.

Hanna, A. D. 1935. Fertility and tolerance of low temperature in *Euchalcidia carybori* Hanna (Hymenoptera: Chalcidinae). Bull. Entomol. Res. 26: 315-322.

Heacox, A. E., R. A. Leopold & J. D. Brammer. 1985. Survival of house fly embryos cooled in the presence of dimethylsulfoxide. Cryo-Lett. 6: 305-312.

Heilmann, L. J., J. D. DeVault, R. A. Leopold & S. K. Narang. 1993. Improvement of natural enemies for biological control: a genetic engineering approach *In* S. K. Narang, A. C. Bartlett & R. M. Faust, eds. *Applications of Genetics to Arthropods of Biological Control Significance.* pp 167-189. CRC Press, Boca Raton, Florida.

Hofsvang, T. & E. B. Hågvar. 1977. Cold storage tolerance and supercooling points of mummies of *Ephedrus cerasicola* Stary and *Aphidius colemani* Viereck (Hym. Aphidiidae). Norweigian J. Entomol. 24: 1-6.

Houseweart, M. W., S. G. Southard & D. T. Jennings. 1982. Availability and acceptability of spruce budworm eggs to parasitism by the egg parasitoid, *Trichogramma minutum* (Hymenoptera: Trichogrammatidae). Can. Entomol. 114: 657-666.

Hu, Z. W. & Q. Y. Xu. 1988. Studies on frozen storage of rice moth and oak silkworm *In* J. Voegele, J. K. Waage & J. C. van Lenteren, eds. *Trichogramma and Other Egg Parasites.* Colloques de l'INRA 43: 327-338.

Huang, M. D., S. H. Mai, W. N. Wu & C. L. Poo. 1974. The bionomics of *Anastatus* sp. and its utilization for the control of lichee stink bug, *Tessaratoma papillosa* Drury. Acta Entomol. Sinica 17: 362-375.

Hunter, C. D. 1994. Suppliers of Beneficial Organisms in North America. California Environ. Prot. Ag., Dept. Pest. Reg. 30 pp.

Iacob, M. & N. Iacob. 1972. Influence of temperature variations on the resistance of *Trichogramma evanescens* Westw. to storage with a view to field release. Analele Institutului de Cercetari pentru Protectia Plantelor 8: 191-199.

Jackson, C. G. 1986. Effects of cold storage of adult *Anaphes ovijentatus* on survival, longevity and oviposition. Southwestern Entomol. 11: 149-153.

Javier, P. A., A. Havron, B. Morallo-Rejesus & D. Rosen. 1991. Selection for pesticide resistance in *Aphytis* III. Male selection. Ent. Exp. Appl. 61: 237-245.

Jayanth, K. P. & S. Nagarkatti. 1985. Low temperature storage of adults of *Bracon brevicornis* Wesmael (Hymenoptera: Braconidae). Entomon 10: 39-41.

King, C. B. R. 1934. Cold storage effect on *Trichogramma* and on eggs of *Ephestia kuehniella*. Tea Quart. 1: 19-27.

Klunker, R. 1982. The host suitability of cold-preserved *Musca domestica* puparia for the mass rearing of the hymenopteran *Muscidifurax raptor* (Hymenoptera: Pteromalidae). Angewwandte Parasitologie 23: 32-42.

Kok, L. T. & T. J. McAvoy. 1983. Refrigeration, a practical technique for storage of eggs of *Trichosirocalus horridus* (Coleoptera: Curculionidae). Can. Entomol. 115: 1537-1538.

Krishnamoorthy, A. 1989. Effect of cold storage on the emergence and survival of the adult exotic parasitoid, *Leptomastix dactylopii* How. (Hym., Encytidae). Entomon 14: 313-318.

Kovalenkov, V. G. & N. V. Kozlova. 1981. Seasonal colonization of *Habrobracon*. Zashchita Rastenii No. 12, 33-34.

Lee, R. E. 1989. Insect cold hardiness: to freeze or not to freeze. Bioscience 39:308-313.

Lee, R. E. 1991. Principals of insect low temperature tolerance, *In* R. E. Lee & D. L. Denlinger, eds. *Insects at Low Temperature*, pp 17- 46. Chapman & Hall, New York.

Legner, E. F. 1976. Low storage temperature effects on the reproductive potential of three parasites of *Musca domestica*. Ann. Entomol. Soc. Amer. 69: 435-441.

Leopold, R. A. 1998. Cold storage of insects: using cryopreservation and dormancy as an aid to mass rearing. *In: Area-Wide Control of Insect Pests Integrating the Sterile Insect and Related Nuclear and Other Techniques*. FAO/IAEA Internat. Conf., Penang, Malaysia. June, 1998. IAEA-CN-71.

Leopold, R. A., R. R. Rojas & P. W. Atkinson. 1998. Post pupariation cold storage of three species of flies: increasing chilling tolerance by acclimation and recurrent recovery periods. Cryobiology 36: 213-224.

Leppla, N. C. 1984. Systems management of insect population suppression programs based on mass-propagation of biological control organisms *In* E. G. King & N. C. Leppla, eds. *Advances and Challenges of Insect Rearing*. pp 292-294. USDA Publ. Washington, D.C.

Liang, M. D. & C. Hu. 1989. Bionomics of *Telenomus theophilae*. Acta Ser. Sinica 15: 95-99.

Lindquist, D. A., M. Abusowa & M. J. R. Hall. 1992. The New World screwworm fly in Libya: a review of its introduction and eradication. Med. & Vet. Entomol. 6; 2-8.

Liu, J. J. & Y. Tian. 1987. Cold storage of *Encarsia formosa* Gahen. Chin. J. Biol Cont. 3: 4-6.

Ma, H. Y. 1986. Studies on long-term storage of hosts for propagating *Trichogramma*. *In* J. Voegele, J. K. Waage & J. C. van Lenteren, eds. *Trichogramma and Other Egg Parasites*. Colloques de l'INRA 43: 369-371.

Ma, W. Y., J. W. Peng & Y. X. Zuo. 1988. Studies on the biology of *Trichogramma dendrolimi* Matsumura. Scientia Silvae Sinicae 24; 488-495.

Mazur, P., K. W. Cole, J. W. Hall, P. D. Schreuders & A. P. Mahowald. 1992a. Cryobiological preservation of *Drosophila* embryos. Science 258: 1932-1935.

Mazur, P., U. Schneider, and A. P. Mahowald. 1992b. Characteristics and kinetics of subzero chilling injury in Drosophila embryos. Cryobiol. 29: 39-68.

McDonald, R. C. & L. T. Kok. 1990. Post refrigeration viability of *Pteromalus puparum* (Hymenoptera: Pteromalidae) within host chrysalids. J. Entomol. Sci. 25: 409-413.

Medina, C. P. & E. P. Cadapan. 1982. Mass rearing of *Corcyra cephalonica* Stn. and *Trichogramma species*. Philippine Entomologist 5: 181-198.

Miles, J. E. & J. S. Bale. 1995. Analysis of chilling injury in the biological control agent *Aphidoletes aphidimyza*. Cryobiol. 32: 436-443.

Mills, N. J. & V. G. Nealis. 1992. European field collections and Canadian releases of *Ceranthia samarensis* (Dipt.: Tachinidae) a parasitoid of the gypsy moth. Entomophaga 37: 181-191.

Morewood, W. D. 1992. Cold storage of *Phytoseiulus persimilis* (Phytosiidae). Exp. Appl. Acarol. 13: 231-236.

Morrison, R. K. 1988. Methods for the long-term storage and utilization of eggs of *Sitotroga cerealella* (Olivier) for production of *Trichogramma pretiosum* Riley.

In J. Voegele, J. K. Waage & J. C. van Lenteren, eds. *Trichogramma and Other Egg Parasites.* Colloques de l'INRA 43: 373-377.

O'Brochta, D. A. & P. W. Atkinson. 1997. Recent developments in transgenic insect technology. Parasitology Today 13: 99-104.

Patel, A. G., D. N. Yadav & R. C. Patel. 1988. Effect of low temperature storage on *Campoletis chlorideae* Uchida (Hymenoptera: Ichneumonidae) an important endo-larval parasite of *Heliothis armigera* Hubner (Lepidoptera: Noctuidae). Gujarat Agricul. Univ. Res. J. 14: 79-80.

Petters, R. M. & D. S. Grosch. 1977. Reproductive perfomance of *Bracon hebeto*r females with more or fewer than normal ovarioles. Ann. Entomol. Soc. Amer. 70: 577-582.

Piao, Y. F., H. Lin & G. R. Shi. 1992. Quality control of the physique of mass-reared *Trichogramma.* Plant Protect. 18: 28-29.

Pizzol, J. & J. Voegele. 1988. The diapause of *Trichogramma maidis* Pintureau & Voegele in relation to some characteristics of its alternative host *Ephestia kuehneilla* Zell. *In* M. Boulétreau & G. Bonnot, eds. *Parasitoid Insects.* Colloques de l'INRA 48: 93-94.

Polgár, L., M. Mackauer & W. Völkl. 1991. Diapause induction in two species of aphid parasitoids: the influence of aphid morph. J. Insect. Physiol. 37: 699-702.

Polgár, L., B. Darvas & W. Völkl. 1995. Induction of dormancy in aphid parasitoids: implications for enhancing their field effectiveness. Agricul. Ecosys. Environ. 52: 19-23.

Pu, T. S., Z. H. Liu & Y. X. Zhang. 1988. Studies on *Trichogramma. In* J. Voegele, J. K. Waage & J. C. van Lenteren, eds. *Trichogramma and Other Egg Parasitoids.* Colloques de l'INRA 43: 551-556.

Ramos, J. P. & J. Jimenez. V. 1993. Cold preservation of eggs of *Sitotroga cerealella* (Olivier) in refrigeration and vacuum packing at different pressures. Rev. Colombiana Entomol. 19: 64-71.

Rudolf, E., J. C. Malausa, P. Millot & R. Pralavario. 1993. Influence of cold temperature on biological characteristics of *Orius laevigatus* and *Orius majusculus* (Het.: Anthocoridae). Entomophaga 38: 317-325.

Salt, R. W. 1936. Studies on the freezing process in insects. Tech. Bull. 116. Univ. Minn. Agric. Exper. Statn.

Salt, R. W. 1953. The influence of food on cold hardiness of insects. Can. Entomol. 85: 261-269.

Schread, J. C. & P. Garman. 1934. Some effects of refrigeration on the biology of *Trichogramma* in artificial breeding. J. New York Entomol. Soc. 42: 268-283.

Salt, R. W. 1959. Role of glycerol in the cold-hardening of *Bracon cephi* (Gahan). Can. J. Zool. 37: 59-69.

Shalaby, F. F. & J. M. Rabasse. 1979. Effect of conservation of the aphid parasite *Aphidius matricariae* Hal. (Hymenoptera; Aphidiidae) on adult longevity, mortality and emergence. Ann. Agricul. Sci. (Moshtohor) 11: 59-73.

266

Shekharappa, K. Jairao & Y. Suhas. 1988. Effect of refrigeration of puparia on adult emergence and pupal period of *Eucelatoria bryani*. Indian J. Agricul. Sci. 58: 875-877.

Shi, Z. H., S. S. Liu, W. L. Xu & J. H. He. 1993. Comparative studies on the biological characteristics of geographic/host populations of *Trichogramma dendrolimi* (Hym.: Trichogrammatidae) in China. III. Response to temperature and humidity. Chin. J. Biol. Cont. 9: 97-101.

Singh, R. & M. Srivastava. 1988. Effect of cold storage of mummies of *Aphis craccivora* Koch subjected to different pre-storage temperature on per cent emergence of *Trioxys indicus* Subba Rao & Sharma. Insect Sci. Application 9: 647-657.

Slovak, M. 1988. Breaking diapause of the ichneumonid *Exetastes cinctipes* by low temperature. Biol. Czech. 43: 549-554.

Sømme, L. 1966. The effect of temperature, anoxia, or injection of various substances on haemolymph composition and supercooling in larvae of *Anagasta kuehniella* (Zell.). J. Insect Physiol. 12: 1069-1083.

Sømme, L. 1995. *Invertebrates in Hot and Cold Arid Environments*. 275 pp. Berlin: Springer-Verlag.

Spradbery, J. P. 1990. *Australian Screwworm Fly Manual of Operations*. CSIRO Aust. Div. Entomol. Tech. Paper No. 49. 241 pp.

Stary, P. 1970. Methods of mass-rearing, collection and release of *Aphidius smithi* (Hymenotpera: Aphidiidae) in Czech. Acta Entomol. Bohem. 67: 339-346.

Steponkus, P. & S. Caldwell. 1993. An optimized procedure for the cryopreservation of *Drosophila melanogaster* embryos. Cryo-Lett. 14: 375-380.

Strong-Gunderson, J. M, R. E. Lee, M. R. Lee, K. S. Grove and T. J. Riga. 1990. Ingestion of ice nucleating bacteria increases the supercooling point of the lady beetle *Hippodamia convergens*. J. Insect Physiol. 36 153-157.

Sun, X. L. & E. Y. Yu. 1988. Use of *Trichogramma dendrolimi* in forest pest control in China *In* J. Voegele, J. K. Waage & J. C. van Lenteren, eds. *Trichogramma and Other Egg Parasitoids*. Colloques de l'INRA 43: 591-596.

Tanahara, A. & S. Kirihara. 1989. Recovery speed of sterilized adults of the melon fly, *Dacus cucurbitae* Coquillett (Diptera: Tephritidae), anaesthetized by chilling and various gases. Jap. J. Appl. Entomol. Zool. 33: 99-101.

Tauber, M. J., M. A. Hoy & D. C. Herzog. 1985. Biological control in agricultural IPM systems: a brief overview of the current status and future prospects. *In* M. A. Hoy & D. C. Herzog, eds. *Biological Control in Agricultural IPM Systems*. pp 3-9. Orlando: Academic Press.

Tauber, M. J., C. A. Tauber & S. Masaki. 1986. *Seasonal Adaptations of Insects*. New York: Oxford University Press.

Tauber, M. J., C. A. Tauber & S. Gardescu. 1993. Prolonged storage of *Chrysoperla carnea* (Neuroptera: Chrysopidae). Environ. Entomol. 22: 843-848.

Temerak, S. A. 1984. Effect of different constant temperatures on certain biological aspects of *Bracon brevicornis* Wemael, a larval ectoparasitoid of *Sesamia cretia* Led. Bull. Soc. Entomol. Egypte No. 63, 129-134.

Tzanakakis, M. E. & D. C. Stamopoulos. 1978. Survival and laying ability of *Dacus oleae* (Diptera: Tephritidae), cold stored as pupae and adults. Z. Angewandte Entomol. 86: 311-314.

van der Linden, A. 1990. Survival of leafminer parasitoids, *Chrysocharis oscinidis* Asmead and *Opius pallipes* Wesmael after cold storage of host pupae. Mededelingen van de Faculteit Landbouwwetenschappen, Rijksuniversiteit Gent. 55: 2a, 355-360.

van Lenteren, J. & J. Woets. 1988. Biological control in greenhouses. Ann. Rev. Entomol. 33: 239-269.

Vigil, B. O. 1971. Laboratory multiplication and release of *Trichogramma sp.* with a view to controlling *Heliothis zea* (Boddie) and *Alabama argillacea* (Hb.) in El Salvador (Central America). Coton Fibres Tropicales 26: 211-216.

Voegele, J., J. Daumal, P. Brun & J. Onillon. 1974. Effect of exposure of the egg of *Ephestia kuehneilla* (Pyralidae) to cold and ultraviolet light on the multiplication rate of *Trichogramma evanescens* and *T. brasiliensis* (Hymenoptera: Trichogrammatidae). Entomophaga 19:341-348.

Wan, F. G. & R. Wang. 1990. The survival rate and fecundity of *Zygogramma suturalis* (Col.: Chrysomelidae) under low temperature. Chin. J. Biol. Cont. 6: 145-147.

Wang, T. & J. E. Liang. 1989. Diapause termination and morphogenesis of *Holcothorax testaceipes* Ratzeburg (Hymenoptera: Ecyrtidae), an introduced parasitoid of the spotted tentiform leafminer, *Phyllonorycter blancardella* (F.) (Lepidoptera: Gracillariidae). Can. Entomolgist 121: 65-74.

Wang, C. L., H. X. Wang, H. Lu, C. M. Gui, G. M. Cheng & D. J. Chui. 1988. Studies on technique of long term keeping host eggs of *Trichogramma* under ultralow temperature *In* J. Voegele, J. K. Waage & J. C. van Lenteren, eds. *Trichogramma and Other Egg Parasitiods*. Colloques l'INRA 43: 399-401.

Whitaker-Deerberg, R. L., G. J. Michels, L. E. Wendel & M. Farooqui. 1994. The effect of short-term cold storage on emergence of *Aphelinus asychis* Walker (Hymenoptera: Aphelinidae) mummies. Southwestern Entomol. 19: 115-118

Young, S. R. & W. Block. 1980. Experimental studies on the cold tolerance of *Alaskozetes antarcticus*. J. Insect Physiol. 26: 189-200.

Zachariassen, K. E. 1985. Physiology of cold tolerance in insects. Physiol. Rev. 65: 799-832.

Zachariassen, K. E. & H. T. Hammel. 1976. Nucleating agents in the haemolymph of insects tolerant to freezing. Nature 262: 285-287.

Zaslavski, V. A. 1988. *Insect Development: Photoperiodic and Temperature Control*. Springer-Verlag, Berlin. 187pp.

Zhang, G. J. 1992. Effect of cold storage on the longevity, sex ratio and reproduction of *Spalangia endius* (Hymenoptera: Pteromalidae). Chin. J. Biol. Cont. 8: 19-21.

Zhu, D. F. & Y. H. Zhang. 1987. Cold storage of *Trichogramma* developed from fluctuating temperature. Natural Enemies of Insects 9: 111-114.

10

Insect Control in the Field Using Temperature Extremes

Temperature extremes limit the geographic range of insect populations, either causing direct mortality or limiting the range of host plants or animals. The extent to which pest managers can make use of temperature extremes to limit the size of insect populations and protect commodities depends upon their ability to manipulate exposure of insect pests to those temperatures. Pest managers can raise temperatures locally in the field, usually with fire. Burning vegetation is an ancient practice with many goals, including insect control. Technology that has become available within the past 100 years makes application of fire specifically for insect pest control feasible.

Although some insects have physiological adaptations for withstanding heat stress, the heat generated by intense solar radiation or open flame is a broad spectrum control measure. Its use must, therefore, be carefully considered in light of total effects on the ecosystem, particularly on plants. The effect of high temperatures on the commodity to be protected frequently limits the opportunities for its use in pest management. Extremely cold temperatures are difficult to generate in the field. Therefore, climatic conditions set limits on the extent to which pest managers can manipulate low temperature extremes. Cold temperature manipulations typically take the form of maximizing exposure of insects to naturally occurring cold temperatures by removing insulation. In this chapter, I discuss some of the considerations and opportunities for using temperature extremes to control insects in the field, primarily crop and natural habitats. Many of the examples used come from my own research on cultural controls for the Colorado potato beetle, *Leptinotarsa decemlineata*. The principles, however, can be considered for any pest system.

Control Using Low Temperature Extremes

The low temperatures experienced in the field are determined by climate and weather patterns. Low temperatures can be manipulated only to a limited extent in the field. Although application of very cold liquids like nitrogen is theoretically feasible, it would be much more costly than using a combustible fuel and high temperatures to achieve the same effect. Options for using low temperatures in pest control in the field also are limited by insect physiological and behavioral mechanisms for avoiding low temperature mortality. Insects do use insulation by overwintering inside host plants, leaf litter, or soil rather than in more exposed sites. The insulating properties of insect overwintering sites either maintain higher temperatures than ambient air or prevent very rapid changes in temperature. Pest managers can manipulate insulation by removing it, for example by altering the location of insects in soil or preventing buildup of snow cover. Either lowering the minimum temperature experienced by the insect or increasing the variation in temperature, producing rapid declines in particular, could increase mortality (see Chapter 3).

Insects avoid low temperature extremes by migration and dispersal or by diapause, in which physiological mechanisms protect against low temperatures. Both responses require timing; insects use cues like photoperiod and host quality, as well as temperature itself, to time either migration or diapause. If the timing of dispersal or diapause can be manipulated strategically, e.g. by manipulating the quality of host plants, opportunities might be found for increased exposure to lethal cold temperatures. The strategy would be to prevent diapause induction or dispersal before the onset of lethal temperatures.

Manipulating Low Temperature Extremes

The soil acts as effective insulation from air temperatures. Temperature moderation in the soil profile is well known in agricultural meteorology (Rosenberg et al. 1983). Crop stubble and plant debris act as major sources of soil insulation. In addition they help to retain snow cover, which acts as additional insulation. Manipulating the low temperature extremes to which insect pests are subjected amounts to removing as much insulation as possible, exposing the insect to as close to ambient temperatures as possible.

One way that soil temperature can be manipulated is by the use of mulch or ground cover. Soil that has been mulched remains at higher temperatures than bare soil (Fig. 10.1). If removed immediately before a rapid drop in

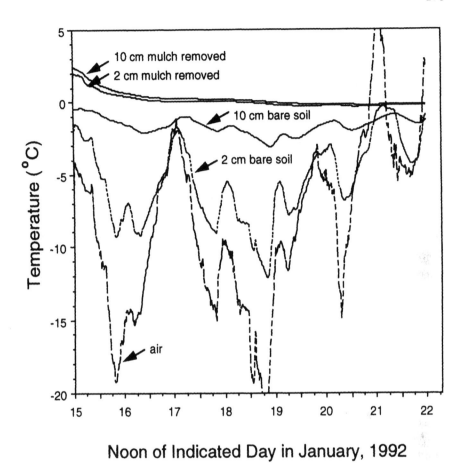

FIGURE 10.1 Air and soil temperature profiles during passage of a cold front, January 15-22, 1992, Wooster, Ohio. Probes were placed at two depths in plots that had been mulched with straw until January 15, 1992 and plots with bare soil since September, 1991 (unpublished data).

temperature is forecasted, the soil temperatures can fall very quickly over a short period of time (Milner et al. 1992). The rapid decrease helps avoid cold temperature conditioning in the insect pest (see Chapter 3). Effects of the ensuing rapid drop in temperature on Colorado potato beetle have been inconsistent. Kung et al. (1992) demonstrated in laboratory experiments that rapid (1 °C/h) declines in temperature from approximately 0 °C to less than −4 °C would cause >90% mortality in overwintering adults. Mulch removal resulted in statistically significant reductions in overwintering Colorado potato beetle survival in Wisconsin during 1991 (Milner et al. 1992) and 1993 (67%, unpublished data). No significant reductions in survival, however, were found during 1992 in Wisconsin or in either 1992 or 1993 in Ohio (unpublished data). Temperature data recorded in Ohio during the days after mulch removal suggested that the higher soil moisture under mulch prevented a temperature drop to lethal lows (Fig. 10.1). Therefore, the mulch and mulch removal technique seems most appropriate for soils that remain dry during winter, even under mulch, such as the relatively sandy soils in Wisconsin.

Tradeoffs in removing insulation, snow cover, and plant debris include water loss and soil erosion. Mulch is not particularly expensive; grass or small grains can be grown on the overwintering site and mowed to mulch the soil. Labor costs for removing the mulch, however, can be high because when covered with snow and frozen into a solid mat, mulch cannot be raked easily. This soil temperature manipulation probably will not be feasible unless used on limited acreage. We have found that overwintering Colorado potato beetle adults can be concentrated in chosen areas by manipulating host plant availability and quality during late summer (Hoy et al. 1996). Compared with other controls, however, including control by burning the beetles before overwintering, manipulating soil temperatures with mulch has not proved economically viable.

Manipulating Spatial Distribution of Insects

The distribution of insects in the soil affects their exposure to cold temperatures. In general, the deeper in the soil the more temperature is moderated (Rosenberg et al. 1983, Fig. 10.1). For the insect, deep overwintering depths are harder to achieve, may become saturated more frequently, and may remain cool and delay emergence when surface conditions do become favorable. These tradeoffs lead to a characteristic distribution of insects in the soil (Fig. 10.2). Location of insects in the soil can be altered by tillage. Movement of plastic beads in the soil profile during moldboard plowing and rigid-time cultivation have been measured (Cousens & Moss 1990), primarily to estimate the effects of tillage on weed seed germination. The transition probabilities estimated by sampling for the plastic beads after tillage can be used to generate estimates of changes in distribution of insects in the soil profile (Fig. 10.2). After fall chisel plowing, a greater proportion of overwintering Colorado potato beetles were found at depths between 0 and 5 cm in plots that had been chisel plowed during the fall than in plots left undisturbed during fall. The altered distribution of beetles was similar to that predicted for rigid tine cultivation (Cousens & Moss 1990). No differences

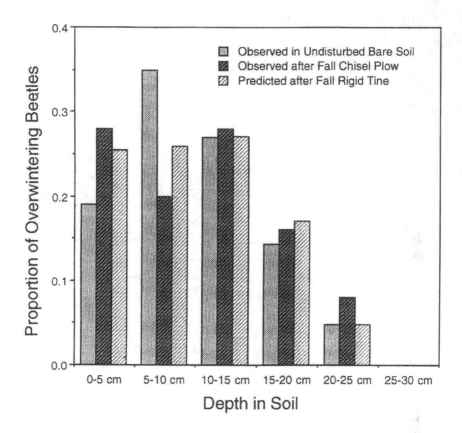

FIGURE 10.2 Distribution in the soil profile of overwintering Colorado potato beetles. Observed distributions were measured in plots in Wooster, Ohio. Predicted distribution was calculated from the observed distribution in undisturbed soil and transition probabilities for rigid tine cultivation estimated by Cousens & Moss (1990).

in overwintering mortality were found in these plots, but the data demonstrate that overwintering insects can be moved predictably within the soil profile by cultivation.

Distribution of insects in crop residue can be manipulated by mowing, chopping, disking, and plowing the residue. The extent to which these operations will increase cold induced mortality is determined by their effects on snow retention and insect exposure. For example, mowing corn stubble may result in greater wind speeds and less snow retention, i.e. less insulation, than leaving the stubble standing.

Manipulating Temporal Distribution of Insects

Phenology of herbivorous insect pests typically is well adapted to exploiting crop plants. Pest managers often consider pest phenology in adjusting planting dates to avoid exposure of the crop to colonizing insect pests and to time monitoring and insecticidal treatments. An often overlooked opportunity to manage pest populations, however, is to slow development of the pest so that is has not reached the overwintering life stage when temperatures and host plant quality begin to decline. Consequently, the insect's adaptations for cold temperatures can be avoided. Late planting and use of *Bacillus thuringiensis* insecticides can delay Colorado potato beetle population development so that much of the second generation will not have reached the adult stage when the crop has senesced. Pupae and adults that have not had a chance to feed before overwintering may be less likely to survive freezing temperatures. In another example, early maturing cotton varieties can be harvested and the crop refuse destroyed before the boll weevil, *Anthonomus grandis,* is physiologically prepared for overwintering (Sturm et al. 1990).

Control Using High Temperature Extremes

High temperature extremes occur naturally and have been generated in field situations with solarization, steam, or fire. Fire is by far the easiest method for generating intense and localized heat. Literature on control of insects with fire has been reviewed by Hardison (1976), Miller (1979), and Warren et al. (1987). Some of the key points described in these reviews follow. Intense heat can be generated by open burning of vegetation or by mechanical production of flame with fuels like gasoline or propane. Open burning is more typically practiced in ecosystems for which fire is a naturally occurring ecological feature. Although many examples of insect control by fire can be cited, most are by open burning in ecosystems that are preadapted to naturally occurring fire. Prescribed open burning techniques have been adapted from wildfire control techniques to control intensity, spread rate, and coverage. Mechanical burning typically is used in crops for which unharvested plant debris must be destroyed and can harbor pests. Spring burning of alfalfa, grasses, and mint to remove the previous year's crop debris before regrowth has been common practice for many years with associated benefits in insect and disease control. Mechanical burning can be limited by risk of open fire in ecosystems that are not adapted to open fires. Targets of control by burning are often diseases and pests other than insects. Burning rice stubble controls not just rice stem borers but also leaf blast, foot rot, stem rot, and rats. Burning dead and senescent plant material in ryegrass pastures in New Zealand eliminates mycotoxin problems caused by a saprophytic fungus. In these situations, the non-insect pests may be more important targets for burning than the insects.

Direct control of insects by fire is greatest for relatively immobile insects remaining on plants or the soil surface. Therefore, despite the consistent physiological effects of high temperatures on insects, fire can be considered to be a somewhat selective insect control. Those insects that remain below ground or are highly mobile are relatively unaffected by the heat of the fire. Post-burning effects depend on habitat and food web modification. Open burning aided by humans was originally used mainly for habitat modification, often for maintaining good grazing lands. Direct control by fire is often secondary to habitat modification in the effects on insects in open burning systems. The obvious reduction in vegetation removes food for most herbivores, but also favors regrowth that increases availability of food for some species. Altered plant architecture in the burned areas can influence insects in various ways. For example, burning can eliminate breeding sites for tsetse flies and questing sites for ticks. Insect predators, however, are often favored by easier hunting in an architecturally simplified habitat for more concentrated prey, although prey also can be reduced in numbers. Detritivores are often favored by increased detritus following fire, but are not favored by the more xeric conditions that typically follow burning. Given the various effects of habitat modification on insects, it is not surprising that numerous examples of increases, no changes, and decreases in population density following burning have been cited for detritivores, herbivores, predators, and parasites. Blanket statements regarding the expected impact of burning for pest control cannot be made, rather the effects could only be predicted from a thorough understanding of the effects on the habitat and the biology of the pest complex.

Naturally Occurring High Temperatures

Solar radiation can generate temperatures that are lethal to insects in the field. In field situations, temperature often interacts with other abiotic and biotic factors to cause mortality. Higher temperatures in water used for flooding to control wireworms in Florida sugarcane result in greater mortality more rapidly (Hall & Cherry 1993). For the linear relationship between percentage mortality and time under flooding, both the intercept and the slope increase with temperature. More rapid loss of oxygen would be expected under the higher water temperatures, so in this case submersion produces mortality but high temperature intensifies the effect. In xeric environments, mortality can result from heat or from desiccation or from a combination of the two (Sterling et al. 1990). Mortality due to heat and desiccation in abscised flower buds (squares) on the soil surface is one of the most important natural sources of mortality for the cotton boll weevil (Sturm et al. 1990).

Allowing or encouraging solar heating and drying of vegetation can be a useful management technique. Shorey et al. (1989) demonstrated that thorough insect control can be achieved in figs if they are exposed to sunlight. Figs begin the drying process on the ground after dropping from the tree when ripe. Figs placed in the sun were uninfested by insects, but 12% of figs placed in shade were infested (primarily Nitidulidae) within 3 days. Figs naturally falling into sunlight contained almost no insects; 31% of those falling into dense shade were infested. Temperatures inside figs drying on soil in sunlight were typically above 50°C when air temperature in the shade averaged 32°C. On the tree, 50% higher insect infestation was found on the shaded north than the sunny south sides of trees. Cultural practices like pruning and manipulating location of ripening figs could be important management tools in this and similar systems.

Artificially Increased Temperatures

Insect control through artificially increased temperatures is typically achieved by fire. Fire can cause mortality directly and by modifying vegetation to lower the habitat's carrying capacity for the pest. For example, burning vegetation results in direct mortality of ticks, but also decreases availability of questing sites, increases solar radiation and ambient temperature, and decreases humidity, all of which lead to additional mortality after the fire (Wilson 1986). In fact, the effects of burning on ticks were similar to the effects of mowing vegetation, suggesting that direct mortality was less important than habitat modification in this case.

Burning crop residue can protect future crops by destroying pests that would otherwise survive there. The extent to which this practice makes sense is largely governed by agronomic factors. Partial burning of green millet and napier grass stalks provided effective reductions of stalk boring larvae that could carry over into the next millet crop in the Sudano-Sahelian zone of West

Africa (Gahukar 1990). In this case, the crop residue could not be completely destroyed because the stalks are used in fencing, roofing, and animal bedding. To maintain their structural integrity, the stalks must be cut when still green and heated on a metal plate placed over a fire to a temperature just high enough to kill the larvae. A somewhat less effective alternative was to bag the stalks in plastic bags for 3 days, during which solar heating would kill the borers. Control was greater than 70% in 4 of 6 burning and 3 of 6 bagging trials.

Burning crop residue in alfalfa and similar crops is intended to provide protection of the crop in the area that has been burned. The species controlled and the duration of protection depend entirely upon the characteristics of the pest, crop, and burning methods. Short term effects of flaming in alfalfa depend upon the insect species and growth stage at which the alfalfa is burned (Schaber & Entz 1988). *Lygus* spp. were reduced for the year of the burn in plots burned before alfalfa growth but not after, whereas alfalfa weevil and pea aphid populations were reduced only when the burn was at 20-25 cm of growth. Long term studies of annual flaming treatments in alfalfa demonstrated protection from *Lygus* spp. and alfalfa plant bug, *Adelphocoris lineolatus* but had little effect on predator populations (Schaber & Entz 1994). In fact, counts of minute pirate bug, *Orius tristicolor*, were enhanced. Flaming had to be timed correctly (after 50 mm of spring growth) to achieve these long term effects.

Burning that produces sufficient direct mortality to insect pests can protect adjacent areas. Area wide (22 square mile) burning of ditch banks in Washington reduced *Myzus persicae* by 51-91% and number of beet western yellows virus infected beets in adjacent fields by 77-84% (Wallis & Turner 1969). Protection of adjacent forest from pine beetles by burning infested trees has been studied in Canadian forests (Stock & Gorley 1989). For inaccessible areas, burning tracts of felled trees resulted in mortality of mountain pine beetle, *Dendroctonus ponderosae*, close to 100%. A high fuel load and consistently intense fire were required, features which were judged to be unsuitable for sites with poor soil quality because of the degradation they cause. Post-fire attack was also heavy in standing timber at the fire boundaries, due either to stress in those trees or attraction of beetles to the fire itself, or both.

Propane or LP gas fueled burners have been used in pest control for many years. The standard components for a propane burning device are a fuel tank, pressure regulator, hose, and nozzle. Refinements include extensions to allow hand burning in hard-to-reach areas (e.g. ditch banks), multiple nozzles on a boom to treat row crops, and ignition systems that allow switching the flame on and off (Fig. 10.3). Two different kinds of burner systems are available. Vapor burning systems draw and burn vapor from the top of the fuel tank. Because the supply of vapor in the tank is limited by vaporization rate of the liquid fuel, these systems have more limited output than the alternative liquid burning systems. Self-vaporizing liquid burners draw liquid from the bottom of the fuel tank and heat it to a vapor immediately before discharge from the nozzle orifice. Typically the liquid is heated to a vapor by piping it over or through the flame. The self-vaporizing burners are typically ignited at low pressures using vapor drawn from the top of the tank. Once the nozzle has been heated, the operator switches to the liquid fuel using either a solenoid or manual switch. Details on construction of propane burning systems for use in row crops have ben published (Moyer et al. 1992) and are available from several manufacturers.

Interest in propane flamers for control of Colorado potato beetle was revived in the past decade when populations developed widespread resistance to available insecticides. The typical strategy was to treat the edges of fields on warm, calm, sunny days during spring to control adult beetles as they arrived and fed on the tops of potato plants (Moyer et al. 1992). Control in field trials was typically comparable to that achieved with insecticides. In 1991 a propane flamer tested in Quebec reduced adult density by 10-30% and egg hatch by 15-30%, respectively (Duchesne et al. 1992). On Long Island, control of adults was 90% and egg hatch was reduced by 27% (Moyer et al. 1992). In neither study were plants less than 20 cm tall injured.

Direct and immediate mortality is due to the brief and rapid rise in temperature as the burner passes. In studies to calibrate a burner near Wooster, Ohio, the peak temperature achieved on bare soil under a passing burner could be manipulated by altering tractor speed, boom height, and propane fueled output (Fig. 10.4). Temperature was measured with a thin thermocouple, exposed just above the soil surface. The temperature peaked within a few seconds of the burner passing, then rapidly returned to close to ambient temperatures within a minute of the burner passing. Colorado potato beetle survival studies conducted during this calibration test indicated that maximum temperatures greater than 80°C resulted in >80% mortality within 72 hr (Fig. 10.5). Additional tests were conducted on small potato plants (4-6 leaves). Peak temperatures averaged 70.3°C in the canopy of these plants. Beetles were released on the plants immediately before burning, collected immediately afterwards, and held on potato plants. Survival, measured by ability to walk off a hot plate, was less than 1% after 5 d. Potato canopy

FIGURE 10.3 Four row propane burner with solenoid switch and adjustable nozzles, constructed from a kit from Flame Systems, Inc., Eau Claire, Wisc. The burner can be used to control overwintering adults at the edges of fields during spring (top) or before they disperse from the field in late summer (bottom).

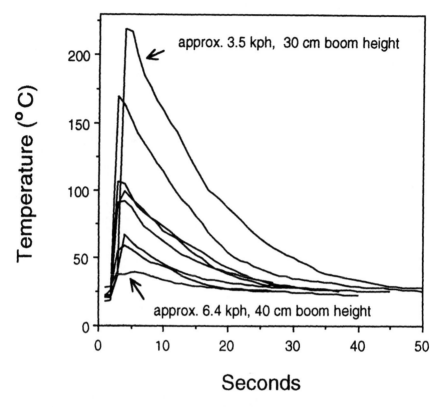

FIGURE 10.4 Temperatures at the soil surface during passage of a four row propane burner. Different profiles were produced by varying tractor speed and boom height.

density, however, reduced the peak temperature attained. The potato crop canopy absorbed enough heat that temperatures and control declined rapidly with increasing plant height. Peak temperatures measured on bare soil and plant heights of 7.5, 15, 23, and 30.5 cm were 120, 75, 68, 50, and 43°C, respectively.

Pelletier et al. (1995) documented injuries caused by propane flaming as loss of muscle function in extremities. They concluded that beetles with flame-damaged tarsi would not be able to climb back onto plants and would eventually starve or dehydrate. Late season flaming can result in further mortality when combined with adverse winter temperatures. Studies in Wayne county, Ohio were conducted with Colorado potato beetles collected before and after August and September flaming in trap crops (Hoy et al. 1996) in each of 5 commercial potato fields. One hundred live beetles were collected before flaming and another hundred, only those still walking, were collected

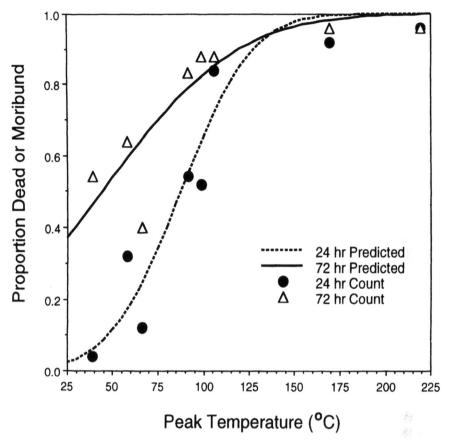

FIGURE 10.5 Colorado potato beetle mortality/moribundity as a function of peak temperature at the soil surface from a passing propane flamer. Beetles were placed on the soil surface immediately before the burner passed. Dead and moribund beetles were counted as those unable to walk off a hot plate in less than 5 minutes. Probit regression was used to generate predicted mortality.

immediately after the flamer had passed. The beetles were caged with potato foliage and tubers within 24 hr of collection; beetles collected before the burning were caged separately from those collected after burning. The soil in the cages was sampled during November, after all live beetles had entered diapause, by digging and inspecting approximately 2.5 cm slices of soil. The distribution of live and dead beetles (Fig. 10.6) confirmed the conclusions of Pelletier (1995) in that a very large percentage of the beetles were found dead at the surface of the soil. In addition, surviving beetles were much more likely to overwinter at shallower depths than beetles that had not been burned, leav-

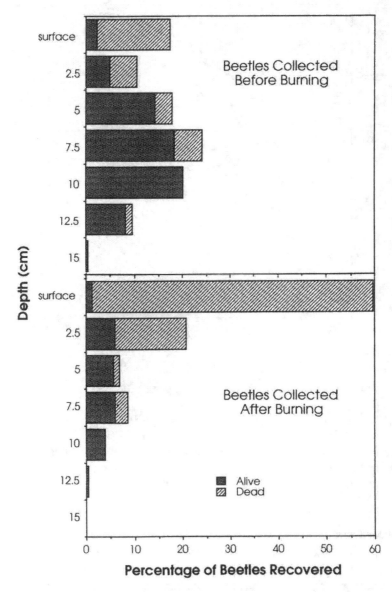

FIGURE 10.6 Percentage of beetles recovered at various depths in soil during November 1994, Wooster, Ohio. Beetles were collected before and after propane flaming in fall trap crops in commercial potato fields, Wayne County, Ohio, and were caged separately on Wooster Silt Loam soil. Each cage contained 100 beetles. Soil was removed in approximately 2.5 cm layers and inspected.

ing them more susceptible to low temperatures during winter. Potato vine killing at crop maturity was very effective with a propane flamer and could be done at speeds up to 9.6 kph (Duchese et al. 1992).

Costs of using propane flame to control insects are comparable to the costs of insecticides, ignoring equipment depreciation. Cost of propane for burning Colorado potato beetle on Long Island, NY, was $12.35/ha compared with insecticides at $61.73/ha (Moyer et al. 1992). Currently, labor costs due to slow speeds required and limited opportunities for use of propane burning equipment (short periods during spring and late summer) are more important than fuel costs as disincentives for potato growers to purchase the equipment for Colorado potato beetle control. Burning is economically viable in other crops. Control of asparagus aphid, *Brachycorynella asparagi*, by burning infested wild asparagus in early spring was estimated to be approximately 1/7 as expensive as either herbicides or digging crowns (Folwell et al. 1990). Aphid control by burning ditch banks cost $47-$115/km from 1964-67 in Washington. Per hectare costs for western beet yellows control in fields surrounding the ditches were $6.17-$16.91. An estimated 3.4-5.2 metric tons of beets/hectare increase in yield, with a value of $55.55-$85.19/hectare, resulted from burning the ditch banks (Wallis & Turner 1969).

Summary and Future Directions

Use of cold temperatures for insect control requires a sufficient understanding of the biology and ecology of the pest to manipulate the insects environment in ways that prevent physiological adaptations or remove insulation. The use of heat to control insects has found niches in particular field situations where crop refuse should be destroyed or the crop ecosystem is adapted to fire. Constraints include cost of generating sufficient heat in field environments, technical ability to generate uniform heating within plant canopies and in the soil, nonselective effects of heat on crops and beneficial organisms, and lack of control for pests with behaviors or spatial distributions that protect them from fire. Given the low cost of propane and solar energy, the first two constraints might be relaxed somewhat by further engineering to improve propane burners or other heat generating equipment and increase the range of uses in crop production for control of weeds and diseases as well as insects. The final constraint is biological and difficult to overcome without the ability to selectively heat insect pests. It will tend to limit the use of heat in the field to niches where pests can be concentrated spatially and exposed at a time when the crop and nontarget organisms are not sensitive to damage. As a general rule, control by high temperature extremes should be considered in situations where pesticide resistance is of concern, crop plants will not be harmed significantly, and insects are immobile enough and remain high enough in the soil or plant profile to be relatively easy targets.

Future possibilities for using heat in the field include lasers and microwave technology as potentially more directed sources of heat. Effects of laser beams on mosquitoes have been measured. Exposing adult *Aedes aegypti* to 130 mw or argon 514.5 mm laser microbeams aimed at the gonadal region for 0.5 sec produced approximately 80% mortality within 24 hr and resulted in no progeny (Rodriguez et al. 1989). Because lasers can interact selectively with various pigments that trap and absorb the radiation, beams that selectively target the pigments in insects and not plants may be possible. With appropriate shielding and given sufficient energy sources, field use of laser heat for insect control may be feasible.

References

Cousens, R. & S. R. Moss. 1990. A model of the effects of cultivation on the vertical distribution of weed seeds within the soil. Weed Research 30: 61-70.

Duchesne, R.-M., D. Bernier & C. Jean. 1992. Utilization of a propane flamer in potato fields in Quebec. Amer. Pot. J. 69: 578.

Folwell, R. J., J. A. Gefre, S. M. Lutz, J. E. Halfhill & G. Tamaki. 1990. Methods and costs of suppressing *Brachycornella asparagi* Mofdvilko (Homoptera: Aphididae). Crop Prot. 9: 259-264.

Gahukar, R. T. 1990. Population ecology of *Acigona ignefusalis* (Lepidoptera: Pyralida) in Senegal. Environ. Entomol. 19: 558-564.

Hall, D. G. & R. H. Cherry. 1993. Effect of temperature in flooding to control the wireworm *Melanotus communis* (Coleoptera: Elateridae). Fla. Entomol. 76: 155-60.

Hardison, J. R. 1976. Fire and flame for plant disease control. Ann. Rev. Phytopathol. 14: 355-379.

Hoy, C. W., J. A. Wyman, T. T. Vaughn, D. A. East & P. Kaufman. 1996. Food, ground cover and Colorado potato beetle (Say) (Coleoptera: Chrysomelidae) dispersal in late summer. J. Econ. Entomol. 89: 963-969.

Kung, K.-J. S., M. Milner, J. A. Wyman, J. Feldman & E. Nordheim. 1992. Survival of colorado potato beetle (Coleoptera: Chrysomelidae) after exposure to subzero thermal shocks during diapause. J. Econ. Entomol. 85: 1695-1700.

Miller, W. E. 1979. Fire as an insect management tool. Bull. Entomol. Soc. Amer. 25: 17-140.

Milner, M., K.-J. S. Kung, J. A. Wyman, J. Feldman & E. Nordheim. 1992. Enhanced overwintering mortality of Colorado potato beetle (Coleoptera: Chrysomelidae) by manipulating the temperature of its diapause habitat. J. Econ. Entomol. 85: 1701-1708.

Moyer, D. D., R. C. Derksen & M. J. McLeod. 1992. Development of a propane flamer for Colorado potato beetle control. Amer. Pot. J. 69: 599.

Pelletier, Y., C. D. McLeon & G. Bernard. 1995. Description of sublethal injuries caused to the Colorado potato beetle (Coleoptera: Chrysomelidae) by propane flamer treatment. J. Econ. Entomol. 88: 1203-1205.

Rodriguez, P. H., W. J. Hamm, F. Garcia, M. Garcia & V. Schirf. 1989. Reduced productivity in adult yellowfever mosquito (Diptera: Culicidae) populations. J. Econ. Entomol. 82: 519-523.

Rosenberg, N. J., B. L. Blad & S. B. Verma. 1983. Microclimate the Biological Environment. Wiley. New York. 495 pp.

Schaber, B. D. & T. Entz. 1988. Effect of spring burning on insects in seed alfalfa fields. J. Econ. Entomol. 81: 668-672.

_____. 1994. Effect of annual and biennial burning of seed alfalfa (lucerne) stubble on populations of lygus (*Lygus* spp.), and alfalfa plant bug (*Adelphocoris lineolatus* (Goeze)) and their predators. Ann. Appl. Biol. 124: 1-9.

Shorey, H. H., L. Ferguson & D. L. Wood. 1989. Solar heating reduces insect infestations in ripening and drying figs. HortScience 24: 443-445.

Sterling, W., A. Dean, A. Hartstack & J. Witz. 1990. Partitioning boll weevil (Coleoptera: Curculionidae) mortality associated with high temperature: Desiccation or thermal death? Environ. Entomol. 19: 1457-1462.

Stock, A. J. & R. A. Gorley. 1989. Observations on a trial of broadcast burning to control an infestation of the mountain pine beetle *Dendroctonus ponderosae*. Can. Entomol. 121: 521-523.

Sturm, M. M., W. L. Sterling & A. W. Hartstack. 1990. Role of natural mortality in boll weevil (Coleoptera: Curculionidae) management programs. J. Econ. Entomol. 83: 1-7.

Wallis, R. L. & J. E. Turner. 1969. Burning weeds in drainage ditches to suppress populations of green peach aphids and incidence of beet western yellows disease in sugarbeets. J. Econ. Entomol. 62: 307-309.

Warren, S. D., C. J. Scifres & P. D. Teel. 1987. Response of grassland arthropods to burning: A review. Agric. Ecosys. Environ. 19: 105-130.

Wilson, M. L. 1986. Reduced abundance of adult *Ixodes dammini* (Acari: Ixodidae) following destruction of vegetation. J. Econ. Entomol. 79: 693-696.

11

Temperature in the Management of Insect and Mite Pests in Greenhouses

Richard K. Lindquist

Worldwide, the total area for greenhouse crop production was estimated by Lenteren & Woets (1988) at 150,000 ha. Only about 18% of this area is covered by traditional glass greenhouse coverings; the remaining area consists of polyethylene or acrylic-covered structures. Some greenhouses have sophisticated environmental control computers that regulate many facets of greenhouse operations, including environmental factors such as temperature, light, and humidification. Others have only minimal environmental controls *e.g.* heaters. The remaining greenhouse area consists largely of polyethylene-covered structures which do not have environmental controls.

Shipp et al. (1991) stated that manipulating the greenhouse environment, including temperature, is perhaps the most underutilized tactic in greenhouse crop pest management. On a practical level, however, there seem to be relatively few options for using temperatures alone to control insects and mites in a greenhouse crop integrated pest management (IPM) program. Most modern greenhouses produce crops more or less continuously. Temperatures are controlled as much as possible to be favorable for plant growth. In most cases these temperatures are also favorable for insect and mite development.

In greenhouses, temperature manipulation, usually in combination with relative humidity/vapor pressure deficits, is used more commonly to manage plant pathogens and weeds than insects and mites. This is mostly because the opportunities are greater; plant pathogens often have quite specific environmental conditions governing their survival. Most insects and mites can survive over quite a wide range of environmental conditions. However, there are some direct and indirect ways that temperatures are or could be used in insect and mite management in greenhouse crops.

Using Temperatures to Control Pests Directly

Steam Treatment of Soils or other Root Media.

Many greenhouse flower and vegetable crops are produced in soil ground beds. Crops such as carnation, chrysanthemum, rose, tomato, cucumber, and lettuce, are commonly grown in this way. Prior to the now widespread use of "soilless" potting mixes, nearly all potted crop plants were also produced in soil. The application of steam heat prior to planting to treat these soil beds, or soil for potted plants, has been done for decades. In addition to killing weed seeds, fungi, bacteria, and most plant viruses, steam treatment can also kill soil insects in areas where temperatures are sufficiently high. The soil temperature must reach 71°C for 30 min following the time that the coolest spot reaches this temperature, to be lethal to insects (Nelson 1991). The soil should be moist for best results. The steam is applied either by using an existing network of perforated pipes buried beneath the soil or is applied over the tops of areas of soil that are covered to retain the steam and heat. Buried steam pipes should be just below the depth of cultivation.

Although quite effective, there are several problems with using steam. Some pests, *e.g.* nematodes, symphylids, and root-feeding insects, can escape to areas deeper within the soil, where the steam does not penetrate. Other, probably more serious problems are cost and time to treat large areas and shortage of equipment. Many greenhouses in which steam pasteurization would be very useful are too large to be treated in a cost-effective way. Unlike North America or northern Europe, greenhouses in Central and South America are not equipped with large boilers for heating. Portable steam generating equipment must be purchased or rented, and this equipment cannot treat large areas at one time.

However, even in areas without built-in equipment, treating plant propagation beds with steam heat using portable steam-generating equipment is a practical pest and disease control method. Propagation areas tend to be on defined benches containing soil or other material that can be easily raised to the proper temperatures throughout.

Soil Solarization and Lethal Warm Air Temperatures

Soil solarization utilizes solar radiation to heat soils. The area to be treated in this way is normally covered with a plastic mulching material with the objective that the sun's rays will generate temperatures high enough to be lethal to pathogens, pests or weeds (Stapelton & DeVay 1995). Nearly all published research deals with results against soil pathogens and weeds (De Vay et al. 1990). Although solarization is successful against some of these organisms, the time involved to complete the process and inconsistent results have prevented widespread use of this technique. However, Stapleton & DeVay (1995) reported that the greatest use of solarization as a pest management tool is in greenhouses, organic farms, and backyard gardens.

The main limiting factors to the use of solarization in greenhouse crop production are the long time that areas must remain out of production (1-2 months), the location of many major greenhouse production areas in cooler climatic areas, and the fact that soil temperatures may not be high enough for good control of many pests. However, it may be possible to combine solarization with reduced dosages of chemical fumigants to obtain good insect control.

In parts of Japan, hundreds of hectares of greenhouses that produce cucumber or sweet pepper are treated by solarization for 7 days following crop removal (Horiuchi 1991). The main target pest for this treatment is the melon thrips, *Thrips palmi*. Heating the soil is not the objective in this case, but rather raising the air temperature to heat the crop residue and the thrips that remain on the plants. All plants are pulled from the soil and left in the greenhouse. This treatment results in significant insect reduction within the greenhouse. However, the long-term effects may be minimal because the thrips are found outdoors on numerous hosts and can easily move into the greenhouses when the next crop is planted, unless insect-screening or some other barriers are used.

Some crop advisors suggest leaving the greenhouse vacant for 1 to 4 wk between crops and maintaining normal plant production temperatures rather than cooling the greenhouse. The theory behind this is that insect and mite pests from the previous crop will continue to develop, emerge from soil or other root media, and starve. This has been documented in commercial greenhouses; for example, Costello & Gillespie (1993) found that maintaining greenhouse temperatures at 25°C for 10 days after removing all plant residues killed adults of the pepper weevil, *Anthonomus eugeni*. However, maintaining this temperature in British Columbia winters was too expensive for profitable crop production, so a combination of maintaining 20°C and deploying large numbers of yellow sticky traps to capture adults was used successfully. Keeping the greenhouse cool (2-10°C) between crops allowed pepper weevils to survive much longer (Table 11.1).

Pepper weevils would likely survive for longer periods without food at higher temperatures than smaller insects such as whiteflies and thrips. For these pest groups the greenhouse might not need to remain vacant for more than a few days if temperatures were warm.

Solarization is only practical in a few situations, even if keeping the greenhouse warm between crops would not be too expensive. Leaving the greenhouse vacant between crops is done more easily with vegetables than ornamentals. Vegetable crops tend to be finished at one time, with greenhouse areas then cleaned and prepared for the subsequent crop. Many greenhouses that produce ornamental crops are never totally empty, either because of the diversity of crops produced or the sequential harvesting and replanting of individual beds or benches of the same crop within a greenhouse. Some specialty crop greenhouses, such as those that produce only bedding plants, or bedding plants in the spring followed by poinsettias in late summer, use this technique by default simply by having large blocks of time between crops.

TABLE 11.1 LT_{50} and LT_{100} in days for adult pepper weevils held without food at different temperatures (modified from Costello & Gillespie 1993).

Temperature (°C)	LT_{50}	LT_{100}
2	11	24
5	13	27
10	12	28
27	4	7

Post-harvest Treatments

One of the most promising areas for the use of temperatures as part of an IPM program for greenhouse ornamentals is in the area of post-harvest disinfestation. Hara (1994) reviewed current practices for post-harvest treatment of ornamental plants. Many greenhouse products, especially cut flowers and foliage plants, are exported from quarantined areas. The presence of insects or mites, whether harmful or beneficial, in flowers or on plants may cause a delay or a rejection of the entire shipment at border quarantine inspection facilities. If flowers must be fumigated as a quarantine treatment, the current fumigants are methyl bromide and hydrogen cyanide. Methyl bromide may not be available in the future due to environmental regulations. Both fumigants can injure flowers, reducing shelf life and/or directly damaging flower parts. Combining warm temperatures with controlled atmospheres and/or methyl bromide fumigation are possibilities to obtain insect and mite control and reduce phytotoxicity (see Chapter 8).

Most of the research on controlled atmospheres in pest management has been with high carbon dioxide or nitrogen concentrations, combined with low oxygen and low temperatures (Hara 1994). For example, one potentially useful method of treating cut flowers is to keep them in an atmosphere containing 30 to 45% CO_2 for one wk at 0-1°C (Seaton & Joyce 1989). Because of the treatment time involved the types of flowers that could be treated are limited to those sent by sea freight. Also, certain tropical flowers are injured at temperatures <10°C.

Temperature also affects the success of fumigation with methyl bromide. Effects are evident on both plants and insects. As a general rule, plant injury increases as temperatures increase. Wit & van de Vrie (1985) found that as treatment temperature increased from 17 to 23°C at 30 gm^{-3} methyl bromide for 1.5 h, phytotoxicity increased in several cut flower crops, including chrysanthemum, carnation, and alstroemeria.

Mortimer & Powell (1984) compared the toxicity of methyl bromide to medium and large leafminer larvae, *Liriomyza trifolii*, on chrysanthemum at 8 and 15°C. As shown in Table 11.2, the percentage mortality of large larvae was affected by fumigation temperature over a range of concentration time products (concentration x exposure time), or CTP's, with less mortality at 8°C.

Vapor heat (hot, water-saturated air) and/or hot water treatments may be useful in certain situations (Hara 1994). Vapor heat (43.3°C for 3h)

TABLE 11.2 Mortality of *L. trifolii* large larvae with methyl bromide fumigation of chrysanthemums at two temperatures and 7 CTP's (Mortimer & Powell 1984); British Crown Copyright, MAFF Central Science Laboratory.

CTP (g h/m³)	15°C % Corrected kill	8°C % Corrected kill
61.6	--	100
54.0	100	96
40.5	100	67
34.0	100	55
27.0	91	26
20.2	69	--
13.5	22	--

offered effective control of bulb flies *Eumerus* sp. on narcissus bulbs. The same temperature for 20-30 min controlled gladiolus thrips, *Thrips simplex*, on gladiolus corms.

Hansen et al. (1992) used vapor heat of 46.6°C and 90-98% RH for 1 h and controlled >90% of several insect pests on a range of tropical flowers. Flowers that were not injured included heliconias, red ginger, and bird of paradise; others, including anthurium and dendrobium orchids, were injured by this treatment.

Dips in hot water at temperatures of 44-49°C for 5-20 min are sometimes used to treat plants that do not tolerate fumigation well. As with fumigation, the effects vary and seem to be similar to the reactions to vapor heat. There are several successful examples of this treatment. Complete control of cockerell scale, *Pseudaulacaspis cockerelli*, was obtained on bird of paradise using hot water dips at 49°C for 5 or 6 min (Hara et al. 1993). Dipping cape jasmine cuttings in 49°C water for 10 min killed more than 99% of green scale, *Coccus viridis*, nymphs and adults (Hara et al. 1994). The same water treatment temperature for 12-15 min killed >95% of the ant *Technomyrmex albipes*, the banana aphid, *Pentalonia nigronervosa*, and the mealybugs *Planococcus citri*, *Pseudococcus affinis*, and *P. longispinus* in red ginger flowers (Hara et al. 1996). There was no phytotoxicity when flowers were conditioned in 39°C air for 2 h prior to water treatment, with shelf life more than twice that of flowers not conditioned.

One of the most important insect problems on greenhouse ornamental crops at the present time is the western flower thrips, *Frankliniella occidentalis*, and another thrips of quarantine significance is the melon thrips, *T. palmi*. Hara et al. (1995) evaluated several postharvest treatments on dendrobium orchids for control of these pests, including insecticide dips and hot water immersion. Dipping flowers in 49.5°C water for 15-20 sec reduced thrips infestations but shortened the vase life of several cultivars.

Hot water treatment is also suggested for certain pests on potted ornamental plants, but this use is very limited and probably applicable only for home hobbyists. Baker (1990) listed immersion for 15 min in 43.5°C water as a control for cyclamen mites, *Phytonemus pallidus*, on African violets and cyclamen.

Indirect Uses of Temperatures

Pest and Biological Control Agent Interactions

It is well-documented that temperatures have significant effects on biological control agent (BCA)-host interactions in greenhouses. Usually temperature and relative humidity/vapor pressure deficit (VPD) combine to affect pest and beneficial insect and mite development. Two well-known examples from past research that illustrate the effects of temperature and moisture are biological control of greenhouse whiteflies, *Trialeurodes vaporariorum*, and two-spotted spider mites, *Tetranychus urticae* .

In one of many studies of whitefly-parasite interactions, Helgesen & Tauber (1974) studied how the relationship between the greenhouse whitefly by the parasitoid *Encarsia formosa* was affected by temperature. The relationship between the two insects was evaluated on a greenhouse poinsettia crop and results showed that the most important factors affecting the success of biological control included maintaining greenhouse temperatures at an average of 23.3°C. Other important factors included the number of parasites introduced and the timing of the introductions. The warmer temperatures favored the parasitoid over the whitefly. Lenteren & Hulspas-Jordaan (1983) stated that the intrinsic rate of increase favored the greenhouse whitefly below 20°C. Above 20°C the intrinsic rate of increase was greater for *Encarsia*.

Temperatures in temperate latitude greenhouses where vegetables or ornamentals are produced may not be in these favorable ranges often enough at critical times to ensure successful parasite establishment, if temperature were the only factor. Using the heating system to ensure that these tempertures are maintained will probably not be economical in winter. Also, temperatures that favor the beneficial organism may not be optimal for crop production. Increasing the number of parasites introduced will help overcome some of the problems with lower temperatures, but here again economic factors are important. More parasites and more frequent releases increase costs.

Another well-researched example illustrating the relationships between temperature and BCA-host interaction is the ability of the phytoseiid predatory mite *Phytoseiulus persimilis* to successfully control two-spotted spider mites (Stenseth 1979). This relationship also is affected by temperature, but in a different manner than the whitefly-*Encarsia* situation described above. Relative humidity/VPD also play significant roles. Basically, the predators do best in high humidity/low VPD situations over a wider range of temperatures between 10 and 30°C, than at low humidity/high VPD. Some commercial insectaries offer a "tank mix" of different phytoseiid predators that will thrive under different environmental conditions.

Current research is directed at further elucidating temperature and moisture interactions for many greenhouse pest-BCA combinations, such as the western flower thrips, *F. occidentalis* and predatory mites such as *Amblyseius cucumeris*. Mathematical models are being developed to help predict the effects of these environmental factors on both pests and beneficials (Shipp & Gillespie 1993, Houten & Lier 1995).

Although these recent and past studies have demonstrated that there are some opportunities to manipulate temperature and moisture in greenhouses to favor BCA'S over the pests, the difficulty for biological control implementation in greenhouses is that temperature will not affect all BCA-pest relationships in a consistent manner. However, knowing what the effects are will help predict and/or explain pest management success or failure in a given crop-pest-biological control system.

Temperature and Diapause Regulation

Some beneficial insects and mites enter diapause during short days in temperate latitudes. Temperature manipulation can be used to help prevent this in some cases, although possibly at too high a cost. Morewood & Gilkeson (1991) found that a predatory mite, *Neoseiulus* (=*Amblyseius*) *cucumeris* did not enter reproductive diapause if greenhouse temperatures remained above 21°C. Maintaining these temperatures in temperate latitude greenhouses during winter is extremely costly.

Temperature and Microbial Pesticides

A number of entomopathogenic fungi have been studied for their effects on insect and mite pests in greenhouses. Fransen (1990) reviewed the literature on fungi affecting whiteflies. Although not yet a commercial product, the fungus *Aschersonia aleyrodis* has been thoroughly researched for its effect on the greenhouse whitefly *T. vaporariorum*. Relative humidity is a very important environmental factor affecting the success or failure of this and other fungi, but temperature can affect results as well. *A. aleyrodis* is able to infect *T. vaporariorum* nymphs at a range of temperatures from 15 to 30°C. Infection takes much longer at 15°C than at 25 or 30°C. The best temperature for development of the fungus is approximately 25°C. Greenhouse whiteflies develop well at 21°C, so it is possible for the whiteflies to develop faster than the fungus and escape infection.

Day-Night Temperature Differences

Differences in day and night temperatures may have implications for biological or chemical control. The relationship between day and night temperature is called DIF, and is the average night temperature subtracted from the average day temperature (Nelson 1991). A positive DIF (warmer day temperatures relative to night temperatures) is the norm. A technique used to control the growth and flowering of some plants without using chemical plant growth retardants is to modify the normal day-night temperature relationship to a negative DIF, *i.e.* cooler temperatures during a portion of the day than night temperatures. The temperature reductions are usually done early in the morning for approximately 2 h before and just after sunrise. Apparently, plants react as if these cooler early morning temperatures are the all day temperatures (Ball 1991). Negative DIF will reduce the height of several plant species and reduce the need for chemical sprays to retard height.

Ascerno & Erwin (1991) conducted experiments to determine effects of 16 positive and negative DIF treatments on greenhouse whitefly development. Results showed no overall differences in whitefly survival, with the number of days to first adult whitefly emergence being equal. However, there were significant differences in whitefly developmental stage distribution. There were fewer adults and more immature stages in treatments with the higher night temperatures. If DIF could be manipulated at practical levels to affect whitefly growth stage development, this could affect the success of biological and/or chemical control. Chemical controls could be timed to affect certain whitefly growth stages and/or biological control introductions could be timed to ensure that susceptible hosts were present.

Future Directions

What is the outlook for using temperatures as insect and mite pest management tactics in greenhouses? It is unlikely that the present primary direct uses of temperatures in greenhouse crop pest management (steam pasteurization, solarization) will increase significantly in the future. In a recent review of integrated pest management in greenhouses, Ramakers & Rabasse (1995) did not mention temperature as a primary or significant pest management tool. However, from the information reviewed for this chapter and the examples presented, combining temperature modification with other biotic and abiotic factors (controlled atmospheres, water immersion, and water vapor) for postharvest treatment seems to offer excellent potential for expanded future use. Developing effective postharvest treatments is important, as methyl bromide, the current primary method, appears to have a limited life in pest management because of impending environmental regulations.

It is also likely that the indirect uses of temperature will increase. As greenhouses install more sophisticated environmental controls and monitoring equipment, knowing how temperatures affect pest development, and pest-BCA interactions will be very useful for predicting outcomes of chemical and biological management programs.

References

Ascerno, M. E. & J. E. Erwin. 1991. Effects of DIF on the greenhouse whitefly. *In* A. D. Ali, ed. *Insect & Disease Management on Ornamentals.* Soc. American Florists. Alexandria, Virginia. pp 116-122.

Baker, J. A. 1990. Greenhouse ornamental and house plant insect control. *In North Carolina Agricultural Chemicals Manual.* College of Agriculture and Life Sciences. N. C. State University. Raleigh, N. Carolina 27695-7603.

Ball, V. 1991. Cool day/warm night technology. *In* V. Ball, ed. *Ball Redbook,* pp 283-287. G. J. Ball Publishing. West Chicago, Illinios.

Costello, R. A. & D. R. Gillespie. 1993. The pepper weevil, *Anthonomus eugenii* Cano as a greenhouse pest in Canada. IOBC Bulletin: Working Group IPM in Glasshouses. 16(2): 31-34.

De Vay, J. E., J. J. Stapleton & C. L. Elmore, eds. 1990. *Soil Solarization,* U. N. Food & Agric. Org. Plant Production and Protec. Paper 109. 396 pp.

Fransen, J. J. 1990. Natural enemies of whiteflies: fungi. *In* D. Gerling, ed. *Whiteflies: Their Bionomics, Pest Status and Management,* pp 187-210. Intercept Ltd. Andover, Hants, United Kingdom.

Hansen, J. D. , A. H. Hara & V. L. Tenbrink. 1992. Vapor heat: a potential treatment to disinfest tropical cut flowers and foliage. HortScience 27: 139-143.

Hara, A. H. 1994. Ornamentals and flowers. *In* R. E. Paull & J. W. Armstrong, eds. *Insect Pests and Fresh Horticultural Products: Treatments and Responses.* CAB International. Wallingford, United Kingdom. pp 329-347.

Hara, A. H., T. Y. Hata, B. K. S. Hu, R. T. Kaneko & V. L. Tenbrink. 1994. Hot-water immersion of cape jasmine cuttings for disinfestation of green scale (Homoptera: Coccidae). J. Econ. Entomol. 87: 1569-1573.

Hara, A. H., T. Y. Hata, B. K. S. Hu & V. L. Tenbrink. 1993. Hot-water immersion as a potential quarantine treatment against *Pseudaulacaspis cockerelli* (Homptera: Diaspididae). J. Econ. Entomol. 86: 1167-1170.

Hara, A. H., T. Y. Hata, V. L. Tenbrink & B. K. S. Hu. 1995. Postharvest treatments against western flower thrips *[Frankliniella occidentalis* (Pergande)] and melon thrips *(Thrips palmi* Karny) on orchids. Ann. Appl. Biol. 126: 403-415.

Hara, A. H., T. Y. Hata, V. L. Tenbrink, B. K. S. Hu & R. T. Kaneko, 1996. Postharvest heat treatment of red ginger flowers as a possible alternative to chemical insecticide dip. Postharvest Biol. & Tech. 7: 137-144.

Helgesen, R. G.& M. J. Tauber. 1974. Biological control of greenhouse whitefly, *Trialeurodes vaporariorum* (Aleyrodidae: Homoptera) on short-term crops by manipulating biotic and abiotic factors. Canadian Entomol. 106: 1175-1188.

Horiuchi, S. 1991. Solarization for greenhouse crops in Japan. *In* J. E. De Vay, J. J. Stapleton & C. L. Elmore, eds. *Soil Solarization,* U. N. Food & Agric. Org. Plant Production and Protec. Paper 109.

Houten, Y. M. van & M. M. van Lier. 1995. Influence of temperature and humidity on the survival of eggs of the thrips predator *Amblyseius cucumeris*. Med. Fac. Landbouww. Rijksuniv. Gent 60: 879-884.

Lenteren, J. C. van & P. M. Hulspas-Jordaan. 1983. Influence of low temperature regimes on the capability of *Encarsia formosa* and other parasites in controlling the greenhouse whitefly, *Trialeurodes vaporariorum*. Bulletin SROP. Working Group Integrated Control in Glasshouses VI(3): 54-70.

Lenteren, J. C. van & J. Woets. 1988. Biological and integrated pest control in greenhouses. Ann. Rev. Entomol. 33: 239-269.

Morewood, W. D. & L. A. Gilkeson. 1991. Diapause induction in the thrips predator *Amblyseius cucumeris* (Acarina: Phytoseiidae) under greenhouse conditions. Entomophaga 36: 253-263.

Mortimer, E. A. & D. F. Powell. 1984. Development of a combined cold storage and methyl bromide fumigation treatment to control the American serpentine leafminer *Liriomyza trifolii* (Diptera: Agromyzidae) in imported chrysanthemum cuttings. Ann. Appl. Biol. 105: 443-454.

Nelson, P. V. 1991. Greenhouse Operation and Management, Edition 4. Prentice-Hall, Englewood Cliffs, New Jersey. 612pp.

Ramakers, P. M. J. & J-M Rabasse. 1995. Integrated pest management in protected cultivation. *In* R. Reuveni, ed. *Novel Approaches to Integrated Pest Management*, pp 199-229. CRC Press. Boca Raton, Florida.

Seaton, K. A. & D. C. Joyce. 1989. Postharvest disinfestation of cut flowers for export. *In: Horticultural Research and Extension Update-1989.* Western Australia Department of Agriculture, South Perth: pp 1-7.

Shipp, J. L., G. J. Boland & L. A. Shaw. 1991. Integrated pest management of disease and arthropod pests of greenhouse vegetable crops in Ontario: current status and future possibilities. Can. J. Plant Sci. 71: 887-914.

Shipp, J. L., & T. J. Gillespie. 1993. Influence of temperature and water vapor pressure deficit on survival of *Frankliniella occidentalis* (Thsyanoptera: Thripidae). Environ. Entomol. 22: 726-732.

Stapleton, J. J. & J. E. DeVay. 1995. Soil solarization: a natural mechanism of integrated pest management. *In* R. Reuveni, ed. *Novel Approaches to Integrated Pest Management*, pp 309-322. CRC Press. Boca Raton, Florida.

Stenseth, C. 1979. Effect of temperature and humidity on the development of *Phytoseiulus persimilis* and its ability to regulate populations of *Tetranychus urticae* (Acarina: Phytoseiidae, Tetranychidae). Entomophaga 24: 311-317.

Wit, A. K. H. & M. van de Vrie. 1985. Fumigation of insects and mites in cutflowers for post harvest control. Meded. Fac. Landbouw. Rijksuniv. Gent. 50: 705-712.

Contributors

Jon P. Costanzo, Department of Zoology, Miami University, Oxford, OH 45056

David L. Denlinger, Department of Entomology, Ohio State University, Columbus, OH 43210

Guy J. Hallman, United States Department of Agriculture, Agricultural Research Service, Weslaco, TX 78596

David J. Horn, Department of Entomology, Ohio State University, Columbus, OH 43210

Casey W. Hoy, Department of Entomology, Ohio State University, Wooster, OH 44691

Marcia R. Lee, Department of Microbiology, Miami University, Oxford, OH 45056

Richard E. Lee, Jr., Department of Zoology, Miami University, Oxford, OH 45056

Roger A. Leopold, United States Department of Agriculture, Agricultural Research Service, Fargo, ND 58105

Richard K. Lindquist, Department of Entomology, Ohio State University, Wooster, OH 44691

Robert L. Mangan, United States Department of Agriculture, Agricultural Research Service, Weslaco, TX 78596

Linda J. Mason, Department of Entomology, Purdue University, West Lafayette, IN 47907

Donald A. Reierson, Department of Entomology, University of California, Riverside, CA 92521

Michael K. Rust, Department of Entomology, University of California, Riverside, CA 92521

C. Allen Strait, Department of Entomology, Purdue University, West Lafayette, IN 47907

George D. Yocum, Department of Entomology, Ohio State University, Columbus, OH 43210

Index

Bold page numbers indicate primary information